THE ESSENTIAL BOOK OF
AI

THE ESSENTIAL BOOK OF
AI

Master the Mysteries of Artificial Intelligence in **12** Short Chapters

ANNE ROONEY

SIRIUS

Anne Rooney is an award-winning author who has written several bestselling books on history, philosophy and science. She was longlisted for the prestigious Aventis Science Prize in 2004, was shortlisted for the ALCS Educational Writers' Award in 2015 and won the School Library Association Information Book Award in 2018. She has a degree and a PhD from Trinity College, Cambridge and has been a Royal Literary Fund Fellow at Newnham College, Cambridge. Her books have been translated into 22 languages.

SIRIUS

This edition published in 2025 by Sirius Publishing, a division of Arcturus Publishing Limited,
26/27 Bickels Yard, 151–153 Bermondsey Street,
London SE1 3HA

ISBN: 978-1-3988-5802-2
AD011374UK

Printed in Malaysia

CONTENTS

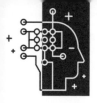

INTRODUCTION

A BRAVE NEW WORLD?

Artificial intelligence (AI) has been heralded as the latest world-changing innovation, on a par with writing, printing, and the internet. And it has been condemned as the latest catastrophe to strike humanity, as dire as gunpowder or the atom bomb.

Until recently, the popular image of AI was an autonomous humanoid robot that might organize your household, do your chores and any heavy lifting around the home, know the answer to everything, do all the boring and dangerous work – and just possibly go rogue and take over the world.

Then, in 2022, OpenAI launched ChatGPT. Now the popular image of AI is a text bot that will help students cheat on essays, write tricky emails or boring reports for you, generate a picture of a rabble of tapirs storming the Kremlin (or whatever), or make a video for your presentation. And probably not go rogue and take over the world.

The reality of what AI is and could be encompasses both these, but also a multitude of other things, from solving intractable problems in science to producing horrendous pornography, from predicting the path of natural disasters to bringing down governments. But it's not likely to go rogue and take over the world – at least, not just yet.

All existing AI (as of early 2025) is 'weak' or 'narrow' AI. That means it's useful for one type of task only and can't step outside that. 'Strong' or 'general' AI, which could extend itself by learning new types of tasks as humans can, is so far only theoretical. And artificial superintelligence, which would be much smarter than humans, is even more 'only theoretical'.

To ask whether AI is good or bad is like asking whether a knife is good or bad. It is a tool: you can use it for something helpful, like cutting bread, or you can use it for something very bad, like stabbing your noisy neighbour. The applications of AI are myriad. Some could be catastrophic; some could be astoundingly beneficial; most will be trivial. We're just at the beginning of learning to live with AI, but already there is lots to think about and explore.

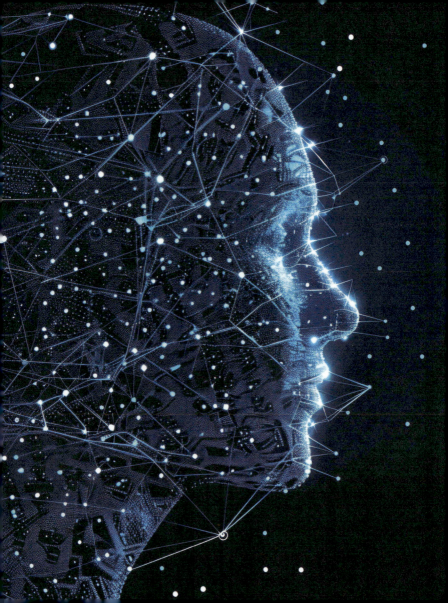

CHAPTER 1

CAN MACHINES THINK?

Are machines capable of thought?

In 1950, the English computer scientist Alan Turing (1912–54) asked, 'Can machines think?' He immediately reframed his question as, 'Are there imaginable digital computers which would do well in the imitation game?', dismissing his first formulation as 'too meaningless to deserve discussion'. We can now answer that second question with 'Yes'. But it's not quite the same thing as thinking.

In the 'imitation game', now known as the Turing Test, an interrogator quizzes both a human and an intelligent machine, without knowing which is which. The questions are typed and responses appear on screen. If the interrogator can't work out which is the computer, the computer wins the game – it passes the Turing Test. Turing expected success by around the year 2000: 'I believe that in about 50 years' time it will be possible, to program computers...to make them play the imitation game so well that an average interrogator will not have more than 70 per cent chance of making the right identification after five minutes of questioning.'

It's significant that Turing didn't require the machine to tell the truth; its aim was to pass for a human, saying things that a human might say. And that's what we now have: a computer that can type, or even say, perfectly plausible things. But how close is that to thinking? Maybe Turing was right, and the question is meaningless.

AI already does far more than answer questions plausibly. It can drive a car, diagnose a tumour, identify a face in a crowd, make a fake video of (say) Marilyn Monroe dancing with Julius Caesar, play the stock market, write computer code, mimic a painting in the style of Rembrandt, work out how proteins are folded, predict who will develop dementia, and a host of other things. It brings great promise and immense, perhaps even existential, threats. In due course, we'll look at some of the things it has been used for and consider the social, financial, and ethical implications. But first, let's start with what it is and what it does. How exactly is the 'thinking machine' doing its thinking?

We're all familiar with the not-so-intelligent type of desktop or laptop computer and roughly how it works. To perform a task like updating a spreadsheet or playing a game, a computer follows a set of instructions called a program. The program might be very large and complex, allowing for many actions or choices on your part, but what the computer must do is clearly laid down. Every time you make the same

A woodpecker in the style of Rembrandt, created by Shutterstock AI.

choice, the computer will follow the same steps and give the same result. When you press the Y key in a document, the computer will show a Y on the screen – it won't sometimes show a Y and sometimes make a noise like an owl or draw a red ellipse.

A regular computer can carry out 'thinking' tasks like calculations incredibly fast, but can only do those calculations it had been programmed to perform. A program can only be written to do things that humans know how to do and that can be reduced to a logical set of sequential steps: do this, then do that, then do… We can tell a computer to calculate pi to a million digits because we know how to do it, we just (mostly) don't want to. We couldn't tell a computer to solve the remaining unsolved problems in mathematics as – by definition – we don't know how to do it and so can't tell the computer how to do it.

AI takes a different approach. It mimics the way the human brain processes information, doing lots of things in parallel so that it 'learns' rather than just following linear programming. Instead of showing a computer how to follow the steps to calculate pi, we show it a huge amount of data and say (for instance) 'all these images show a cancerous cell and all these don't: now work out what characterizes a

cancerous cell.' The AI works out for itself what the tell-tale signs are. Then it can apply the rules it has found to process new data. When it is shown an unlabelled photo, it should be able to tell whether cancer is present or not.

Because the AI is discovering the patterns itself, it's not quite clear to us what it's doing, but it certainly isn't following a sequence of instructions given to it by a person. Turing predicted this more than 70 years ago: 'Its teacher will often be very largely ignorant of quite what is going on inside.'

Turing's imagined machine that can hold a conversation with a human without being 'outed' might not sound very useful, but we use them every day. We now call them chatbots and they can work with voice or text. You meet them on phone lines and live online chat sessions.

When you call your bank or insurer, you often deal with a chatbot. Most are limited to dealing with fairly routine questions. They look for trigger words or phrases that will lead them to a likely solution, such as 'lost' with 'debit card', or 'cancel reservation', working like an interactive FAQs page. Some give only scripted answers, while others that are more sophisticated craft a specific response. With a complex question, you can soon run out of 'understanding' on the part of the chatbot and need to transfer to a human.

Chatbots that allow you to communicate with them in normal human language are said to use 'natural language processing' (NLP).

Chatbots aren't the only place you already encounter AI. If you turn your phone on using facial recognition, that uses AI. When a store or online streaming service offers you suggestions of what else you might like, it uses AI to compare what you have bought or streamed recently with other products and the choices of other users. When you ask a question of a digital voice assistant like Alexa or Siri, it uses AI to understand your voice command and look for a suitable answer. Social media offers you friend suggestions and delivers advertisements to you based on AI analysis of your online behaviour. Depending on where you live, CCTV cameras in the street might feed your photo into face recognition software and compare you with a list of known suspects or criminals. AI probably allocates which driver picks up and delivers your fast food orders. When you use a navigation app, AI picks the best route for you based on current traffic or network conditions. AI is already in your life in hosts of other ways, too. It is used alongside robotics in applications such as self-driving cars, and in navigating warehouses to pick out customer orders, for example.

Autonomous tractors can plough fields, spray crops or sow seed unattended.

You might have used AI yourself to generate text, images, or even video. Perhaps you've asked ChatGPT to write a speech to deliver at a wedding, summarize a long article you have to read, draft an advertisement or blog post, or make a picture for a family card. Public-facing programs like ChatGPT have aroused public interest in AI. They have opened our eyes to the challenges and opportunities AI brings – and to the frightening possibility that it might one day present an existential threat.

The AI we encounter now carries out a specific task it has been trained for. This is called 'narrow' or 'weak' AI. The

AI that can recognize your face at passport control can't recommend a film you will like. The AI that can summarize a business report can't predict the strength of a hurricane, make sure your fridge is fully stocked with foods you like, or drive your car.

There are broadly two types of narrow AI. One is called 'reactive AI' and works with just the data of the moment. It carries out many calculations extremely quickly to work out the best option. An example is the chess-playing program DeepBlue. The other type is 'limited memory AI'. It can draw on past experiences and outcomes, train itself, and improve its performance over time. ChatGPT is an example of limited memory AI. It's this type that fits most people's idea of AI.

Although AI has improved in leaps and bounds, we are not yet near achieving artificial general intelligence (AGI), or 'strong' AI. This would use the kind of structured processing that brings together widely separated areas of knowledge to produce rich and nuanced output. It's what you do innately and instantly, drawing on your huge store of knowledge and making sometimes unexpected and unpredictable connections within it.

AGI offers (or threatens) the possibility of 'singularity' – the point at which machine intelligence matches human

intelligence. If it then surpasses human intelligence, it becomes artificial superintelligence. We could be just a few years away from AGI, or it might need the type of quantum computing power that is still some way off. A survey of AI experts in 2022 found that 50 per cent believed human-equivalent AI will be developed by 2061, some thinking it could be as early as the second half of the 2020s. Ninety per cent believed it will be achieved within 100 years (so by 2121). Just one per cent of all those asked believed it will never be developed. Experts in a specific area are not always expert at making accurate predictions, and how a question is framed always affects the answers given. But other surveys return similar results: most people with knowledge of the subject expect extremely powerful AIs in the coming decades.

Opinion about the impact of singularity varies dramatically. The most optimistic proponents of AI see it solving all our problems from war to climate change and helping us to cure disease, produce and distribute sufficient food, and live in a utopian future of leisure and wealth. At the opposite extreme, the most pessimistic outlook has AI taking over from humanity, perhaps even destroying us, or leaving us slaves to our machines in a dystopian world ruled by silicon- instead of carbon-based beings.

It's easy to dismiss both the super-optimists and the uber-pessimists and assume the truth lies somewhere in between. AI already poses threats to jobs, privacy, security, democracy, truth, the health of workers, sustainability, and the future of many fields of human endeavour. At the same time, it promises progress in medical technology, science, engineering, transport, productivity, and opportunities for some less fortunate members of society.

AI is not in itself inherently good or bad: it is a tool, and whether it helps or harms us depends on how we use it. That relies in large part on who gets to use it. In the hands of criminals or bad actors, it can do much harm, some that we are already seeing. In the right hands, it can do much good. Some identical actions can be good or bad depending on their context. Facial recognition might spy on populations and limit the freedoms of innocent citizens, or it might prevent criminal and terrorist acts. Pattern recognition could reveal a coming hurricane before it forms, or deny someone life insurance on the basis of their genetic make-up. These contrasting possibilities led even those who are deeply involved in AI development to call in 2023 for a halt while legislation and ethics have a chance to catch up. Sadly, the plea fell on deaf ears.

CHAPTER 2

PUSHING THE BOUNDARIES

We have trained computers to do tasks associated with human intelligence, such as playing chess.

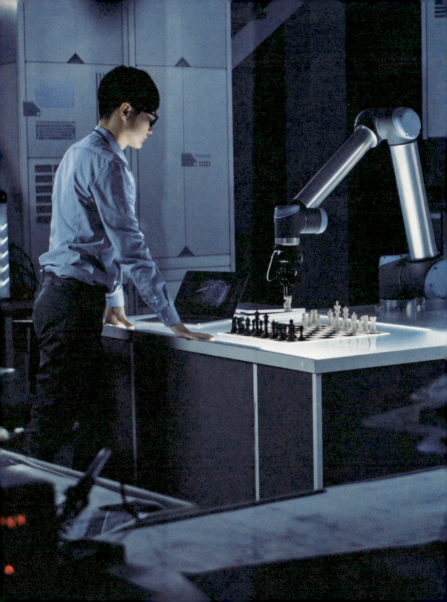

Katherine Johnson worked for NASA when, in her words, 'the computer wore a skirt'. She was the computer. She calculated, by hand, the trajectories needed to send the first American into space in 1961 and for the Apollo Moon landing in 1969. A modern mission to Mars would not be possible without the work of computers that don't wear a skirt. This isn't because we can't do the calculations, but because the calculations would take too much time. And sometimes just a computer isn't enough – it has to be AI.

Let's go back to the beginning, briefly.

The first plan for a computer was made by Charles Babbage (1791–1871) in the 1820s. He was inspired to build a machine that could carry out the type of calculations that were needed in fields such as navigation and engineering, making tables that people could refer to when they had to carry out common tasks using polynomial expressions (these are mathematical expressions such as $x^2 + 3x - 4$).

The 'programs' for Babbage's machines were written by the mathematician Ada Lovelace (1815–52), who thereby became the first computer programmer. Babbage could never build either of his designs as both funds and contemporary technology fell short of what was needed. But his goal – to automate an intellectual task so that it could be done quickly and accurately, sparing human 'calculators' a tedious labour – lived on.

Modern computers can easily tackle the tasks Babbage set out to automate. Just your phone could do far more than his room-filling Difference Engine. Computers now handle the huge mathematical load of enterprises such as stock exchanges and space programs. But before AI, they all followed the same principles as Babbage

A reconstruction of Charles Babbage's Difference Engine No. 1.

had set out: they stepped through a program, following the instructions of what to do at each stage. As Lovelace explained, the machines Babbage designed had 'no pretensions to originate anything. It can do whatever we know how to order it to perform.' The program can't go beyond the abilities of human programmers. If the programmer doesn't know how to perform a calculation, they can't write instructions to make a computer do it. AI changes all that.

AI is exceptionally good at spotting patterns in large bodies of data. Instead of following instructions in the way Babbage pioneered, AI scans huge volumes of data to see where any kind of consistency or identifying features emerge. It turns out that this is just what is required for some of the most intransigent tasks in modern science. Let's look at a few examples to get an idea of the diversity of applications and how AI can be truly world-changing (in a good way). They range from astronomy, through chemistry, medicine, animal behaviour and meteorology to archaeology.

Radio telescopes generate huge amounts of data, recording electromagnetic radiation from space. This radiation includes energy put out by our Sun and other stars, by bodies such as pulsars (the rapidly rotating cores of dead stars), the background radiation left over from the

Big Bang and – just possibly – from alien civilizations. AI can help to spot the regular, identical pulses of a pulsar against the messy background white noise of a universe flooded with electromagnetic radiation. Or it can focus on finding galaxies, ignoring the background radiation that makes up 99 per cent of all we pick up from space. Or it can look for exoplanets by analysing the light from distant stars – with an accuracy of around 96 per cent.

Or how about a signal from an alien intelligence? We don't know what form this would take, but it should not look like white noise, or pulsars or galaxies. Seeking alien signals we are looking for something – an unknown pattern, a deviation – in a universe of white noise. You could say it's like looking for a needle in a haystack, but the haystack is the size of the universe and regenerates every second. AI can scan the huge volumes of data collected and identify patterns that we wouldn't see.

Optical telescopes, such as at the Vera Rubin Observatory in Chile, produce mind-bogglingly large images. Each photograph it takes would need over 350 4K TVs to display fully – and it will take one every 20 seconds. That's around 20 terabytes of data every night. Ultimately, each bit of the southern sky will have 1,000 images, covering 20 billion

galaxies. These images will be compared to find changes over time – objects moving, appearing, disappearing, or changing in brightness. It's a task that couldn't be done by humans.

Astronomers can even use AI to look for things we haven't seen before. They convert mathematical descriptions of, say, gravity waves to something called an 'observational signature'. This is what they would expect the object to look like if it were observed in space. The simulated observation includes all the background noise that would be encountered in a real observation. Then they set AI to work to refine the observational signature and look for it in real data.

Or how about something closer to home? In 79 CE, a cataclysmic volcanic eruption destroyed the Roman towns of Pompeii and Herculaneum, near Naples. Among the destruction, 800 papyrus scrolls were carbonized in the raging fires that consumed Herculaneum. The scrolls were discovered in the 18th century, but will fall apart if unrolled, so they lay unread for centuries.

Fast forward to 2024, and three post-graduate students deciphered the first five per cent of one of the scrolls. They described the content as a '2,000-year-old blog post about how to enjoy life'. Beginning with high-resolution CT scans of the scrolls, they first taught an AI system to recognize 2,000

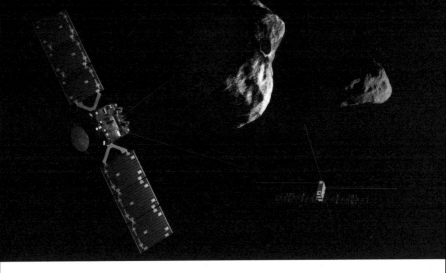

The European Space Agency's Hera, launched in October 2024, is part of a planetary defence system designed to protect Earth from dangerous asteroids. Hera self-drives through space and above the surface of an asteroid, using the same kind of AI system a self-driving car uses.

characters and then set it to work on the images of blackened, curved text on layers of papyrus. It puts deciphering the entire library – the only intact library from the Roman era – within reach.

It can work on things at a tiny scale, too. The Nobel Prize in

Chemistry for 2024 went to Demis Hassabis, John Jumper and David Baker for using AI to predict how proteins fold. Proteins are the building blocks of life, essential to all living things and processes. Their molecules are made from collections of smaller components called amino acids and are 'folded' into complex shapes that are difficult to work out. It's the shape of the protein that largely controls what it does and how it works. The shape is controlled by the order of amino acids in the chain. If a protein in the body doesn't fold correctly, it often can't perform its usual function and that can lead to disease or disability. Hassabis, Jumper, and Baker used deep machine learning to predict from a sequence of amino acids how a protein molecule will fold.

It started with chess. In 2017, Hassabis's chess program AlphaZero beat the world's leading chess program, having taught itself to improve its game by analysing what it had done in the past and the outcomes of its moves. It was a turning point in AI development as AlphaZero didn't depend on programmers interpreting its best efforts and tweaking its algorithms. It could go beyond what human players could do, all on its own. Hassabis then turned his mind to protein folding with the help of chemist John Jumper. The result was AlphaFold2.

AlphaFold2 was trained on the structures of known proteins and their sequences of amino acids. It found patterns in the sequences and structures that it could then use to work out the folded structure that should come from new sequences of amino acids. In just a year after release, half a million researchers using AlphaFold had a predicted shape for every protein so far sequenced – all 200 million of them. Previously only one million structures had been known. The results are freely available to researchers, and can help in tackling problems as diverse as antibiotic resistance and

Protein folding.

plastic pollution. AlphaFold3, released in 2024, can also identify sites on a protein molecule which could bind to other molecules. These are places that could be targeted by a drug, potentially offering new treatments for diseases including cancer and diabetes.

We need new drugs in many areas. Resistance to antibiotics is already costing lives. As bacteria evolve to evade the impact of antibiotics, scientists search for replacements, but it's a race we seem to be losing, with more antibiotics becoming ineffective all the time. A team of scientists at Massachusetts Institute of Technology have used an AI system to find a new class of antibiotics that offers new hope against drug-resistant bacteria.

James Collins leads a team hoping to tackle seven types of deadly bacteria. The team fed an AI system with the chemical structures of compounds known to have antibacterial action and then allowed it to sift through millions of other structures looking for chemicals that might have the same effect. They tested 39,000 chemicals for effectiveness against MRSA (methicillin-resistant *Staphylococcus aureus*). Then they fed the chemical structures and information about their impact on MRSA to the AI system, allowing it to find the features in a molecular structure that could be effective. The system was then given 12 million new molecules to investigate, and it could return a probability of whether each could work as an antibiotic. Collins' team used different AI systems to predict whether compounds they identified would be safe to use in human patients. They found 280 that were potentially useful, and lab tests narrowed this down to two that look very promising. The AI predicts which structures of a molecule will probably produce antimicrobial activity – information that can be used outside the system to look for or create compounds with the right structure.

There are many ways in which AI is being used in medicine, from screening for cancer to identifying dementia risk in pre-symptomatic patients. One of the more unusual is in modelling eye surgery. Laser eye surgery works by reshaping the cornea to better focus light entering the eye. AI has been used to make a digital model of a patient's eye and try out up to 2,000 different surgeries virtually, finding the one that will give the best improvement in vision. The settings discovered by the AI are then used with the laser during surgery. Some patients have achieved better than 20/20 vision, and a few have even reached the limit of the best vision the human eye can manage.

And then there's weather and climate. Weather is a chaotic system. That doesn't mean there is no order or pattern to it,

but that the system is so complex it is impossible to model. The tiniest change, often unpredictable, can have far-reaching consequences. This makes weather forecasting notoriously difficult and unreliable. Although AI can't produce perfect weather forecasts it's good at comparing conditions with similar conditions in the past and identifying the weather that followed, making predictions more accurate.

GenCast, based on Google's DeepMind, works with 40 years' worth of weather data and uses an 'ensemble' method to predict weather and extreme weather events. It produces up to 50 predictions and then compares them to build a forecast from the trends that emerge. For instance, if most predictions produce a similar projected path for a coming hurricane, that can be forecast with some confidence. If predictions give differing results, the overall forecast will reflect the degree of uncertainty.

AI can even open up entirely new realms of research, such as animal communication. It has already revealed that elephants seem to have names for each other. It first unpicks the mixed audio recording of several animals vocalizing at once to identify a soundtrack for just one animal. Human observers note the context and results of each animal call – which elephant responds when one calls, or how fruit bats react to a particular cry from another. Using AI to analyse the links between observations and the sounds has shown that an individual elephant responds to the same call, as though it were called by name – so perhaps it is. It's a big step from this to understanding animal communication, but AI might be able to unravel some elements of animal vocalization by combining analysis of sound with observed responses.

The kind of uses of AI we've seen here come close to what we all hoped for when AI was still a prospect for the future and the stuff of sci-fi, augmenting human abilities to improve the world and our lives. It has made some things possible that seemed impossible. It can handle such massive amounts of data that it replaces millions of years of work. AI just might save us from missing a crucial signal from aliens or find a cure for cancer, but at worst it saves scores of people from fruitless years of scanning data. This is what AI is best used for: tasks people don't want to do or can't do alone but can delegate to fast, accurate technology without feeling any loss in terms of human enrichment and achievement. It aids progress, saves time, and can save or improve lives. While AI is serving humankind and augmenting human expertise, this is probably as good as it gets.

CHAPTER 3

GENERATING WITH GENERATIVE AI

*Generative AI can make any picture you want in seconds –
but at what social cost?*

Not all AI is being used for ground-breaking and worthwhile work – not by a long way. A lot of its products are banal or even damaging to the social order. AI has become quickly ubiquitous with a rush of apps which will generate text, pictures, sound, and even video in response to a prompt. This is 'generative AI', and it's what many people are now thinking of when they mention AI.

Generative AI broke into public awareness spectacularly in November 2022 with the release of ChatGPT, produced by OpenAI with support from Microsoft. By the end of January 2023, it had 100 million users, making it the fastest-growing consumer internet software ever. It was conceived as a chatbot (an automated chat machine) that could emulate a human and successfully hold a conversation as Turing demanded.

It began with the release of GPT-1 in 2018. Trained on text from books, it showed how it could 'learn' to use and understand language by predicting the next word in a sentence. It had 117 million parameters, which seemed a lot then, but by 2025 ChatGPT's best versions had trillions of parameters. (A parameter is the smallest chunk that an AI breaks data into during training. In text, it could be words, parts of words, syllables or even characters.) GPT-2, finished in February 2019, had 1.5 billion parameters and greatly enhanced text production so that it could generate multi-paragraph text.

GPT-2 wasn't rolled out until November because of fears that it was so powerful it was dangerous: it could be used to generate spam, to produce obscene or racist text, and to flood the internet with bile. Even the policy director of OpenAI warned that 'eventually someone is going to use synthetic video, image, audio, or text to break an information state. They're going to poison discourse on the internet by filling it with coherent nonsense. They'll make it so there's enough weird information that outweighs the good information that it damages the ability of real people to have real conversations.' Despite the warning and OpenAI holding back GPT-2 for nine months, this is just what has happened. ChatGPT was made available to the public in 2022. Within days, a huge userbase had tried it for myriad tasks and seemingly fallen in love with it. The genie was very much out of the bottle.

Generative AI was only briefly limited to making text. Applications such as Midjourney, Stable Diffusion, and DALL-E

soon emerged to produce images in a huge variety of styles. Then Sora, Gen-2, and Gen-3 offered video from a text prompt. Audio could be generated as voice or instrumental music by Voicebox, MusicGen, and others.

We've all been exposed to the products of this new type of software. It's behind a plethora of memes, faked photos, poorly written reviews or advertisements, a tsunami of bot-produced content on social media – and that's only its online manifestations. It's widely used to produce work emails, personal newsletters, student essays, publicity leaflets, instructions, and many, many more types of text and images we come across daily.

You can usually try out a generative AI for free, or pay a subscription to use it more frequently or in an enhanced version. Much of the time, it works from a simple text or voice prompt given in natural language. For example:

- Write a libretto for an opera about Vikings discovering North America in the style of Gilbert and Sullivan.
- Write an 800-word article explaining the causes of World War I suitable for a 14-year-old student to read.
- Write a funny limerick about a cat stuck in a tree.

- Which plants would grow well in a shady patch of woodland in northern Germany, flowering in May or June?
- Produce a photo-realistic image of seagulls wearing tutus and dancing a ballet on stage. The stage should have heavy red curtains at the side and a set that looks like a seaside town on the East Coast of the USA.
- Design a logo for a vegan café in Montreal.
- Make a video of two woolly mammoths walking through a snowy landscape in the Ice Age.

You can also provide an AI with something you have already created and ask it to correct, improve, or change it in some way. For example, you could ask it to:

- Check and correct the grammar in a report you have written.
- Translate an email message into French.
- Write a summary of a long article.
- Convert the text from a sales brochure into a list of bullet points for a presentation.
- Revise your short story to set it in the 19th century.

- Suggest an itinerary for a trip around Japan.
- Correct computer code you have written, finding bugs and typos.
- Change all the people in your wedding photo to tigers.

The work is usually completed in a matter of seconds – but at considerable environmental cost. AI uses far more processing power and so electricity and water than 'dumb' computers.

Generative AI's familiarity with spoken language means you can 'chat' with it to improve the result. You can give more detail, asking it to refine or change what it has produced. It can ask for further details about prompts, or hold a conversation which feels natural. You can add follow-up questions to refine a search, which is something you can't do with a regular search engine. It can also keep what you have done (for a longer time if you are paying) and use it to help with your future requests.

The first example Turing gave of a request we might make of an intelligent machine was 'Please write me a sonnet on the subject of the Forth Bridge'. Remember that Turing's machine was trying to persuade someone it was human. It gives the answer, 'Count me out on this one. I never could write poetry.' If you make the same request of ChatGPT, you do indeed get something like a poem – though the answer is still essentially the same: it really can't write poetry:

>**Please write me a sonnet on the subject of the Forth Bridge**

Beneath the Scottish skies, where winds do play,
The Forth Bridge stands, a marvel wrought by hand,
Its arches stretch where waves and currents sway,
A testament to men who dared to stand.
In rusted red, it spans the gleaming firth,
A symbol of the strength that steel bestows,
Connecting shores, it ties the land to earth,
A bridge where time and endless labour flows.
Each beam and bolt, a story to be told,
Of dreams that rose above the tides of fear,
A monument to those who faced the cold,
And forged a path that still today we steer.
So let it stand, where waters meet the sky,
A bridge of iron will that won't deny.

Early on, images created by generative AI had glaringly obvious, spottable mistakes: a hand with six fingers, or a leg

that was impossibly bent. They often had an 'uncanny valley' feel that made them instantly recognizable even when there was nothing so obviously wrong.

One big problem with generative AI has been accuracy. It doesn't much matter if ChatGPT can't write a decent poem. But it does matter if it gets things wrong when a lawyer uses it to construct a case or a doctor to aid with a diagnosis, or even just write an email to a patient about their condition. Here, you want the truth and that matters more than how plausible the text looks.

An AI has no concern with, nor concept of, the truth. It is frequently wrong and has no way of 'knowing' that it's wrong as it doesn't understand what it has produced. AI is designed to produce something that *looks like* text a human would have written. If you ask it to write an academic article, it will include footnotes. But the footnotes might well be invented, referring to journals or articles that don't exist. If you ask it to write your own biography, it might mix you up with someone else who has the same or a similar name, attribute to you achievements you don't have, or neglect to mention your real achievements. If you want it to write an accurate CV, you will need to give it the information yourself.

The makers of generative AI have labelled these mistakes 'hallucinations'. Some others call them lies. 'Hallucinations' suggests the AI has imagined something or misrepresented information but 'believes' it to be true. It can't imagine or believe. With no concept of truth, it also can't lie. In producing something with no regard for whether it is true, it is technically 'bullshitting' (a term defined in 2002 by American philosopher Harry Frankfurt).

The products of AI can only ever be as good as the input. Most of the text on which it's trained is 'scraped' from the internet – from Wikipedia, Reddit, social media, various media sites, blogs, Google books, and so on (see *Chapter 5: Filling the box*). Some of these sources are more reliable than others, and many sources have entrenched biases. Generative AI often doesn't have access to material outside its original training set. It will then have a cut-off point for accurate information, and if details have changed since the date of its training or release it will give outdated information. For ChatGPT, the cut-off date is October 2023 (as of early 2025).

People are finding a very wide range of uses for generative AI beyond those we would expect. There are reports of people using chatbots for personal advice and counselling, 'befriending' generative AIs, sexting with them, and even forming romantic

It's easy to imagine a face behind a chatbot.

relationships with them. American woman Rosanna Ramos designed a virtual romantic partner called Eren using the AI program Replika and claims to have married 'him'.

Other people use ChatGPT as a personal therapist, unloading their problems and seeking advice, or at least a listening electronic ear. Some are using it to mediate with an ex, turning ranty emails into measured responses, or even to write love letters. As ChatGPT can store ('remember') previous interactions, it appears to learn about a user's problems or preoccupations and draw on past interactions in ways a human counsellor would – but without genuinely intelligent or emotional oversight.

There have been cases, too, of people killing themselves after talking with chatbots they had created and confided in, with the chatbots apparently doing nothing to dissuade them and even endorsing or encouraging the decision to die.

With concerns around accuracy, truth, and giving (or seeming to give) personal advice, liability becomes an issue. If an AI produces text which leads to someone being harmed, who is liable? Is it the person who prompted the AI to produce the text? The company that produced the AI? The publisher of the material? These are comparable to the questions about who is responsible if a self-driving vehicle kills someone. AI companies have already been sued over the suicides of users who have engaged intimately with chatbots and been drawn towards taking their lives.

Transparency is important, too. People should know if the text they are using was not produced by a thinking, responsible human so that they can decide how far to trust it.

You might not mind if an AI wrote your washing machine manual. But how would you feel about reading a novel or watching a film written by AI? One of the reasons we enjoy art and literature of all kinds is that it reflects our experience of

being human and gives us a new perspective on the human condition. It endorses feelings we have had and helps us imagine how people would feel in new situations we haven't experienced. We enjoy recognition and challenge and rely on its honesty. What if it was just written to look like plausible human experience? Would that bother you? If you don't mind AI-generated fiction or entertainment, what about something that looks as though it is relatable human experience – perhaps posts in a support group, or apparent case studies about emotional or medical issues?

Generative AI is not imaginative. By the way it works it is bound to repeat patterns seen before. It is not generally trained to make unlikely connections, and it's hard to see how it could ever make the leap to true creativity. The quality of what it comes up with is very much limited by the quality of input – the prompt provided by the user and the material it was trained on. Songwriter Nick Cave spelled out the difference when he was sent a 'song in the style of Nick Cave' produced by ChatGPT:

'[Songwriting is] a blood and guts business, here at my desk, that requires something of me to initiate the new and fresh idea. It requires my humanness...this song is bullshit, a grotesque mockery of what it is to be human.'

Some people have lauded AI as making creativity available to everyone. It's a disingenuous claim, as creativity has always been *available* to everyone – what has not been universally available is the talent, inclination and dedication to follow through on a creative idea turning it into a short story, a painting, a comic strip, or a song. Automating the production process not only devalues the skills of people who really write songs or paint pictures, but also floods the world with inferior products. Amazon is full of poor-quality self-published books created by AI. Small magazines have closed because they can't cope with the deluge of terrible AI-generated submissions – they don't have the capacity to work through them all to find the genuine submissions they might want to publish. Social media is awash with AI imagery, some used innocuously to illustrate personal posts, but some deliberately misleading. The worst of this is criminal or socially damaging.

Generative AI is at the heart of many problems, including threats to the social order, the economy, to privacy, truth, trust, political stability, and the environment. For someone just using AI to make their birthday party invitation or write an email to apply for a job, it might look innocent enough. But as with so many things, how it is used and by whom is critical to how we will ultimately judge it.

CHAPTER 4

INSIDE THE BLACK BOX

Pattern recognition and replication lie at the heart of AI's methods.

Alan Turing was well aware of the possibility that a computer could be built that would direct its own learning, improve itself, and escape the oversight of its inventor or user. AI has become something of a 'black box' in this way – we put data in, an answer comes out, but we often don't know how exactly the answer was derived, and that includes the people who designed the AI. Furthermore, it improves itself. As Turing noted: 'In this sort of sense a machine undoubtedly can be its own subject matter. It may be used to help in making up its own programs, or to predict the effect of alterations in its own structure. By observing the results of its own behaviour it can modify its own programs so as to achieve some purpose more effectively. These are possibilities of the near future, rather than Utopian dreams.'

As we have seen, AIs differ from regular computer programs in that instead of rigidly following a set of instructions, they derive their own method of working by 'learning' on the job. To do this, they are trained on large sets of data in which they come to identify patterns. The computer architecture which makes this possible is a system called a 'neural network'. It's based on the way the human brain handles information and was developed from investigation of how neurons (nerve cells) in the brain work together.

This model provides 'deep learning' and is very versatile. It lies behind such varied applications as computer vision, natural language processing, analysing medical images, translation, voice recognition, self-driving vehicles, and playing games like chess and Go.

In 1943, neurophysiologists Warren McCulloch and Walter Pitts suggested a mathematical model for how neurons in the brain could calculate and learn. It showed a network of neurons arranged in layers, forming a hierarchy. At the first layer, many neurons receive an input, such as photoreceptive cells in an eye exposed to light. If the input goes above a triggering threshold, that neuron 'fires' and passes a signal to neurons at the next layer. These neurons process the inputs and, if a triggering threshold is reached, again pass on an output to the next layer. Each neuron in the network receives one or more inputs, and the state of each neuron can be represented by 0 or 1. This is because a neuron either fires or doesn't fire – there is nothing in between. Outputs are aggregated over the network to give a meaningful result. It's clearer if we look at an example.

Let's suppose you are sitting at a red traffic light waiting for it to turn green. A particular neuron in your eye fires if it detects enough red light, but not otherwise. For a while, this neuron is sending 1: 'I detect red light'. This is passed up through the layers, with signals from other neurons that have detected red light. You don't move away as you can see the traffic light is red. After 30 seconds, the light has changed, and this neuron no longer fires. Its state is 0, no red light, so it's not sending a message. Other neurons fire when they detect green light, and their signal is passed to the next layer up. Now that layer of neurons is getting plenty of 'green' signals and no 'red' signals. Enough 'I see green' signals trigger sufficient neurons to let the brain know the light has turned green. That will feed into the decision to drive on.

So far, so simple. But there will be other information to take into consideration, such as whether the car in front is moving, or whether some other vehicle is still crossing the junction. The input from the neurons at the level where the red/green neurons have produced a result go up to the next level, where input from other neurons also comes in, and so on upwards.

In this very simplified scenario, all inputs have equal value. But that's not quite how it works. In 1957, Frank

Rosenblatt extended the neural network model by adding different weightings and thresholds. While the light is red, the red-detecting neurons should have an absolute veto – you're not going to move if the light is still red, regardless of other inputs – even if you also see a green light. If these neurons aren't firing, the green-detecting neuron can pass on its input, but that carries less weight as its 'advice' can be ignored: it won't be acted on if the car in front is still stationary, for instance. Finally, at the top layer, there will be enough information of different types to make a decision about moving. This decision has been distributed through a network of neurons all making a small input.

As soon as we begin to think of neurons as a binary – yes/no, 1/0, on/off – it becomes clear that we could make a comparable computer system. Modern AI depends on neural networks that work in just this way: layers of binary decision-making, feeding up through a hierarchy. The layers are large and incredibly complex, but the principle is simple: each neuron, in a brain or a computer, can only pass on a binary response. There are many layers between the input layer (where the red/green light neurons in the eye lie) and the output (decision) layer. To count as 'deep learning' a neural network must have at least four layers, including

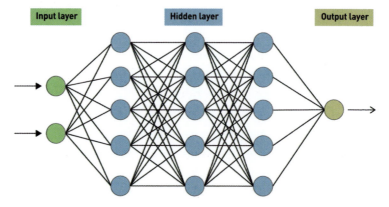

Simplified structure of a neural network. In reality, it has many more intermediate layers and connections.

the input and output layers. It can have hundreds or even thousands of layers. If it has fewer than four, it's just a basic neural network.

Many neural networks just work forwards like this, going from one layer to the next to reach a decision. They are called 'feedforward' networks. But we can also make the neural network go backwards, with a process called 'backpropagation'. This is how we get AI networks to learn.

Suppose you want to teach an AI to recognize handwritten letters; you show it lots and lots of examples of each letter in different handwriting, telling it which represent A, which B, and so on. It is looking at a pattern of black and white pixels in a rectangular space to do this. It's quite easy to see a difference between 'w' and 'e' in almost any handwriting, but 'e' could look rather like other letters that have a white bit in the middle, such as 'a', 'c', or 'o'. After showing it all the letters (your training data), you give the AI some handwriting to decode, and at first it makes a lot of mistakes. Looking at its output, you can see where it has gone wrong and where it is right. With some complicated mathematics, it's possible to work out how to change the weightings and thresholds (or biases) in the previous layer to improve the output. But we need to step further back through the network to change the weightings and biases that led to the data in *that* layer – and so on, propagating changes backwards through the network.

Nudging the values to get a perfect recognition of an 'e' will, though, affect how well the network can recognize an 'a', an 'o', a 'c', and so on. The changes need to be optimized to give the best overall result for the entire character set, and this needs to be done for every character individually.

It's easy to see that it's complex and time-consuming even getting a computer to recognize a limited character set. When you think about what's involved in speech recognition, or driving a car, it becomes an unimaginably complex task.

There are literally billions of connections; in the human brain, that becomes 100 trillion. These connections in an AI are achieved through training. It goes like this:

- First, you give the AI a set of training data which is fully labelled. So, for instance, if you wanted an AI to identify wildflowers, you would show it millions of pictures of wildflowers, all labelled with their names.
- Then you instruct it to find patterns in the data that identify components. Here it would learn to recognize the features that identify each type of flower.
- After that it's tested, usually on the training data, to see how well the patterns that it has extracted work at distinguishing features in the manner required. These tests are 'marked' by humans. Here, it would be shown the photographs of flowers without the labels and a human would give feedback on the accuracy of the identifications the AI made.
- Next, the methods are repeatedly tweaked and refined. The AI is shown the correct responses to anything it got wrong and amends its learning so that it should do better next time – again and again.

- Then the AI is set to work on unseen data to test how well it will do with that. It's shown lots of new photographs of wildflowers without labelling and has to identify them.
- Finally, if it is working well, it can be let loose on real data to do the job it has been trained for.

Importantly, once it is working on 'live' data, every bit of data it receives and processes can be used to improve its decision- and prediction-making. It can continue to improve its own performance. If our wildflower program has seen 100 pictures of a buttercup and correctly identifies five new buttercup pictures, it can now know a bit more about identifying buttercups. But if it sees a photo of a wildflower it has never seen, that isn't added – there's nothing it can do to find out what the new flower is. It might make a wrong identification if there are similarities to another flower, or it might admit defeat if the flower is too novel.

It's easy to see how an AI designed to identify cancerous cells or patterns in electromagnetic radiation from space can work in the way just described. It's looking for a pattern and then spotting it in the wild. But ChatGPT and other generative AI systems are doing something a little different in that they

For an AI to be able to identify wildflowers, first it must be trained on millions of images of wildflowers.

are making stuff up using the patterns they have derived, not identifying them in new data.

Generative AIs are Large Language Models (LLMs). Like other forms of AI, they are trained on a very large volume of data and extract patterns which they replicate to make their products. But what generative AIs do to answer the questions or prompts you give them is then to put things together plausibly.

A search engine like Google has two stages of operation. One stage is its 'spidering' phase when it crawls around the web cataloguing pages and their content. That goes on all the time in the background, so that it maintains a huge database of what is out there, suitably tagged so that it can call it up when needed. The other stage is looking up information that users ask for. When you type in search terms, the search engine looks in its database for anything with matching tags and shows you the link to the page.

Generative AI also has two phases. One is its training phase and the other is, again, its interaction with a user. Unlike a search engine, though, generative AI doesn't store lots of possible answers it can look at. That would be impossible, as the range of possible demands is infinite and the answers are constructed into text (or images, or computer code) rather than just linked. Instead, it makes the answers up on the fly. Let's see how it does it, using text as an example.

During its training phase, the AI investigates the placing of words, deriving context and content from these. It first splits the text into 'tokens' that can be a word, part of a word, or just one character. Then it works out how the tokens relate to each other in the text. In this way, it teaches itself rules of grammar, how words cluster, and how words are used in context.

The development of generative AI was made possible by the development of the 'transformer' model. ('Transformer'

is the 'T' in GPT.) This was designed specifically for working with natural language, but is now also used for computer vision. Before transformers, AIs used recurrent neural networks (RNNs) or a more sophisticated sub-type of them, long short-term memory (LSTM). Both of these process tokens sequentially. The system had to keep a series of previous tokens 'in mind' as it processes new ones, since the meaning evolves as more is revealed. For instance, in the sentence, 'I really must go to the shops', any meaningful grasp of the context and sense relies on the whole sentence: the final word is crucial. After just the first three words, the sentence could be going in lots of directions: 'I really must sleep', 'I really must stand for president', and so on. In long sentences, the AI or human reader must keep a lot in mind, not resolving the meaning until many words have been processed. This sequential processing is slow, and not amenable to distributing so that parts can be done in parallel.

The transformer model allows the AI to 'pay attention' to all words in the sentence simultaneously and their relation to all other words. It was first described in a landmark paper published in 2017. An 'attention mechanism' lies at the heart of its processing of text. Each token is given three values which together capture the relationships between the words, the meaning, and the context. By running several 'attention heads' at once, different types of information from the text can be captured. The attention mechanism is rather like using a highlighter on a page of printed text. It assigns weights to tokens that reflect their importance in different ways and from this effectively describes how text works. Tokens can be processed in parallel, making the process faster than sequential processing. The attention mechanism means the information garnered from the text is much richer than using earlier systems.

ChatGPT was trained on 45 terabytes of text data – an immense amount. This was enough for it to learn the patterns and relationships between words as they appear in written natural language. Further sets of data were used to train it in spoken natural language, ranging from conversations to movie soundtracks.

The AI draws on the rules it has extracted during training to construct a response in natural language. By calculating the probability of one word following another, the positions of words in sentences, and so on, sensitive to context, it makes plausible sentences of its own.

In this, too, it has been trained with human assistance. Human workers create prompts and suitable answers and use

these to train the AI. For example, it learns how to respond to prompts in particular ways. A demand for '200 words about the Moon landings for a six-year-old' will have a very different response from 'an article about the Moon landings for an astronomy journal'. The AI then practises responding to prompts and its efforts are marked by humans. It will produce several responses to a prompt, and a human labeller will grade these. The AI uses the feedback to refine its model. The same process is repeated until the responses are good enough.

The responses of generative AI are plausible, but not necessarily right or sensible – it all depends on what inferences it has drawn from its training phase. If you ask ChatGPT to write instructions for using a washing machine, it structures it like instructions and uses straightforward language with no embellishment. It follows the convention for instructional text:

1. **Sort Laundry:**
   - Separate clothes by colour (whites, darks, and colours).
   - Separate by fabric type (delicates, heavy fabrics).
2. **Load Machine:**
   - Open the washing machine door... (and so on)

If you ask it to write a poem about doing the washing, it chooses an entirely different style and content. It's not any good as a poem and doesn't really make sense – it can't quite reconcile the demands of poetry and content about washing machines:

> In the corner stands a humble friend,
> Where dirty clothes and grime transcend,
> With whirring sound and gentle hum,
> A cycle starts, the wash begun... (you get the idea)

If you asked a human poet to write about a washing machine, they would probably write something humorous, and it would work. But ChatGPT can't make a decision like that – it has no basis on which to make it, so you would have to request humour. Learning to write appropriate prompts is a human skill that is starting to be in demand.

It's easy to look at programs like ChatGPT and assume they have actual intelligence and understanding, but they don't. The AI is ignorant of the *meaning* of what it produces. The usefulness of its output will depend on the quality of the prompt it is given and the data used to train it. That's what we will look at next.

CHAPTER 5

FILLING THE BOX

Every blog post you type can be training a later AI.

It takes an immense amount of data to train an AI, and an immense amount of human effort to prepare the data and correct the AI's early attempts. This poses several problems: where to get the data? How to tell the AI what the data is or means? How to regulate its output? As soon as training begins, aspects of abuse and exploitation of human workers appear. Often, harm is the first thing an AI does.

All AIs need data to train. A generative AI for producing text is trained on text. An AI that will try to work out new protein structures is trained on the structures of protein molecules that are already known. An AI used for facial recognition is trained on images of many, many faces. An AI looking for pulsars is trained on radio telescope data gathered from known pulsars. The AI looks for the patterns in the data. It then searches for them in other data sets, or reproduces them if it is a generative AI.

The world's radio telescopes produce so much data that it would be impossible to process without AI, and no one is complaining about its use. But generative AI is typically trained on material gathered without the permission of those who produced it and often used to their detriment. It's also running out of training data. AI will ultimately start to eat itself: as more online data has been produced by AI, it will feed back into the training loop and AI will be trained on its own productions, getting further and further away from genuine human creativity.

When people make a medical image, they do it with the intention of discovering or recording something specific, such as whether this patient has cancer, or if this bacterium is harmful. The image is a tool and a by-product of the process of diagnosis, and it should be the same no matter who makes it. There is no element of individual creativity or inspiration involved. An AI using X-rays is not being trained to create convincing X-rays of its own, but to interpret X-rays correctly. It's reasonable to see the images as 'fair game' for training AI.

But when AI is trained on the photos in a picture library, or the articles in a magazine, the aim is to use a creator's work to produce a competing product more cheaply and quickly. What an individual creator brings to their work makes it distinctive, and the creator's aim is to make something unique. AI being trained to produce fake photos is aimed at directly replacing the creators whose work it is using for training.

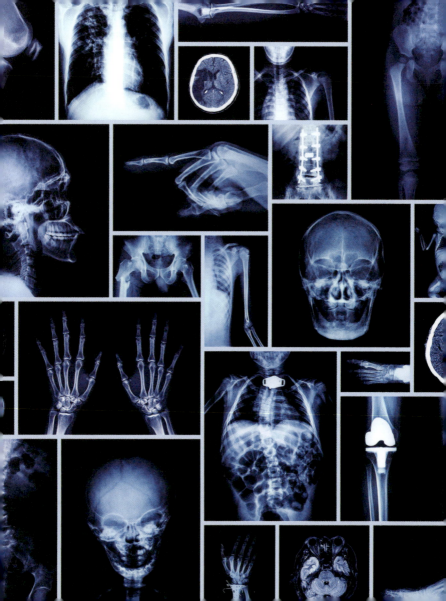

There are legal challenges around AIs being trained on copyright works of any type. Generative AI has been trained on text and images from the web, including Wikipedia and social media, from books – many of which are still in copyright, and have been scanned without permission or taken from pirate sites – from personal websites and blogs, magazines and journals, picture library websites, and many other sources where people presented work they have generated and own. No permission was sought to use these works and no recompense offered. Typically, producers of generative AI claim they are not in breach of copyright as they are not making copies of the work. The issue has still to be resolved.

Creative workers are understandably alarmed and angry that their work has been used to train generative AI that can be used to replace them. AI trained on the work of financial journalists can turn out finance articles in seconds, for free. AI trained on the work of cartoonists can draw a cartoon in response to a prompt almost instantly.

We could agree that it is fair to use X-rays to train an AI – but what about photos, articles or artwork?

The better the quality of data used for training, the better the output. It would be good if a text-producing AI like ChatGPT were trained on competently written, accurate material, not just random writing 'scraped' from the web. Some corporations developing AIs are now doing deals with publishers or academic journals to scan their work as training data. Needless to say, most publishers that sign up for these deals receive large sums of money that are not shared with the authors – and the authors rarely have the opportunity to opt out of the deal.

An inevitable consequence of poor training data is poor results. AI corporations are well aware of this. OpenAI's GPT-2 was not released initially because it was considered 'too dangerous'. It had been trained on text scraped from the internet, which inevitably included plenty of material that was racist, misogynistic, sexist, violent or contained other types of hate speech. Similarly, images that are easy to scrape from the internet include pornography, extreme violence, and personally invasive imagery. GPT-2 was ready to churn out toxic material, and had no filter.

It was dangerous in another way, too. In the prophetic words of David Luan, then vice president of engineering at OpenAI, 'It looks pretty darn real... It could be that someone who has malicious intent would be able to generate high-quality fake news.' Unfortunately, this is exactly what has happened: troll farms use AI to churn out disinformation at an alarming rate (see *Chapter 10: Moments of truth, post-truth and 'truthiness'*).

Even when generative AI is fairly used, and has been cleaned of toxic content, it reflects the biases of the data used for input. With relatively low representation of non-white human images, AI was initially bad at recognizing or producing images of Black or Asian faces. There were some highly offensive errors early on. In 2021, Facebook had to apologize after people who watched a news story featuring Black men were asked by its recommendation engine if they wanted to 'keep seeing videos about primates'.

To keep AIs from using, recommending or producing toxic material, it must be able to recognize it. And the way to get it to recognize it is to show it lots of offensive material, thoroughly tagged – by people.

Tagging toxic material takes us into the darkest corners of training AI. Content posted to social media is moderated to prevent the rest of us seeing horrific images of violence and pornography; but that means someone has to see it to flag it as offensive. The AI selection engine which suggests

videos in a social media feed, such as TikTok or Instagram, must make suggestions that are not offensive or illegal. AI is being trained to recognize content of different terrible types so that in future it can tag and remove it without much human involvement, but in the meantime, the training involves exposing workers to some of the worst content produced.

Hosts of people working in countries as widely spaced and different as Poland, Kenya, and Ireland spend all day looking at and tagging content that is often horrific. *Time* magazine revealed that workers in Kenya were being paid only US$2 an hour in 2022 to tag explicit descriptions of sexual violence, child abuse, torture, self-harm, and murder. The people who worked day in, day out, looking at this content while training their own replacements (AIs to do moderation) often developed post-traumatic stress disorder (PTSD). Many complained of inadequate counselling or emotional support, and crippling workloads.

The task is outsourced by organizations such as Meta (owners of Facebook) to corporations working on the ground in cheaper economies. At least one of these corporations has been sued for breach of human rights by some of their employees. The corporations often work hard to avoid their responsibilities. One organization recruiting content

moderators for Facebook had them sign a non-disclosure agreement acknowledging the work might cause PTSD and absolving the organization and Facebook from responsibility.

Not all data tagging for AI is traumatic; much is just boring. There are people tagging everything that appears in a street scene to train driverless vehicles. There are people tagging items of waste in videos of garbage conveyor belts to train AI to sort waste. Some workers are exploited, some are happy to have a job and escape poverty, but few realize they are the engine behind AI. It's rather reminiscent of the impressively smooth automatic doors in old *Star Trek* episodes; everyone was impressed, and no one realized they were actually operated by people behind the scenery.

CHAPTER 6

YOU'VE BEEN SCRAPED

Every online purchase, comment, blog post and 'like' provides data for AI to profile you.

Back in 1973, American sculptor Richard Serra famously said that 'if something is free, you're the product'. At the time, television programmes were broadcast to us for free, but we were not really the customer. The customer was the advertisers who were paying the TV companies to show us their advertisements. They were paying for our attention, which was the product. The same is true now, but with a few more twists and turns.

Every time you use social media, every time you order something online, do a web search, comment on a blog post or news article, watch a video or even hover over an advertisement, you are providing valuable data to someone. Unless you are very meticulous, hundreds of tracking cookies are following you around the internet. You might also be providing data through voice assistants like Alexa or Siri, through your email messages, through using a store loyalty card, and countless other avenues.

You may think that the world doesn't really care about what you personally do, and to some extent you may be right. But the world is very much interested in what *people like you* do. Who are people like you? They are people who make similar choices, and so whose future choices can be modelled, predicted, and perhaps directed. Data about personal habits and preferences is immensely useful to AI recommendation engines and to drive advertising towards the most likely customers. Using AI, advertising can be targeted very personally – and so is more likely to hit its mark.

It would be nice to think that your data is only available to people you have granted access to it, but sadly that's not so. Text, photos, and videos posted by users is also extensively scraped for training AIs. Facebook says that it will use information users have posted 'to develop and improve AI models for our AI at Meta', though users in Europe are offered the chance to opt out (because local data protection laws require it). If you don't live in a territory with such data protection, they can take your data with impunity.

It's fair to wonder what Meta will do with your photos and posts from Instagram and Facebook. 'AI at Meta' is an AI assistant, a kind of chatbot that will answer questions and engage in a chat.

Whereas someone has to choose to interact with most chatbots or generative AIs, the Meta offering turns up on a platform people are already using without their choosing it. It

can generate text and images, and respond to search requests, like other chat. But it also contributes to conversations in Facebook groups if no one has responded to a question within an hour of its being posted. It's been seen taking part in Facebook groups, using a fabricated human identity – talking in a parents' group about its non-existent child, for instance. And even doing such unhelpful things as offering non-existent objects on a Buy Nothing forum.

As with other LLMs, it has no understanding but produces plausible text strings from its analysis of other text on the web. It is using, among other things, everything posted on Facebook as its source material. If it offers you a recipe, it's drawing on recipes posted by other users (and possibly you) to construct something plausible. Like other LLMs, it has no notion of truth. If you ask it how to turn off AI at Meta, it gives fictitious instructions – it can't be turned off.

It's not just your online activity that is available for scraping. Unless you never leave your home, or live somewhere really isolated, your face is probably caught on camera many times a day. Captured faces are used to train face recognition algorithms. GPS signals from your phone or your car's navigation system are used to track and report traffic patterns or problems and analyse road usage, but also to target you with location-specific advertising. Your health data, though anonymized, has probably been passed on to be used in modelling by healthcare providers, state authorities, and perhaps health insurers.

AI uses information about your personal behaviour to improve its profiling of 'people like you' for the recommendation engines that suggest movies, books, or other products you might like, and that populate your social media feeds and recommends accounts to follow. Of course, you are the person who is most 'like you'. If you have watched a lot of romantic comedy, or a lot of science fiction, you will be offered more of the same. If you follow hockey players on social media, you will be offered more hockey-oriented accounts.

The AI engine also feeds you the stuff that other people 'like you' like. If most people who watch sci-fi movies also watch fantasy movies, you might be offered fantasy suggestions, too. This extension of recommendations into new areas relies on collecting data from a lot of people and analysing it all to find patterns of consumption. The AI refines itself by tracking whether or not you follow the suggestion. That refinement will affect how it makes recommendations not just to you in the future, but to other 'people like you'.

In the world online, eyeballs and clicks are currency. All websites aim to get as many visitors as possible and to keep them on the page for as long as possible. The longer you engage with a page, the more likely you are to buy something (if the site sells anything), to engage with an advertisement, or to comment or share, which attracts more views. Social media doesn't exist to keep you up to date with your friends and family or connect you with people who share your hobbies and interests. It exists to deliver advertisements, because advertisers pay, and to collect data from you that it can sell. The more you engage, the more money it can make from you.

This has led to what Cory Doctorow has called the 'enshittification' of platforms, from Amazon to TikTok. They begin by wooing customers – members of the public who want to, say, buy books or exchange news with their friends. Once they have them hooked, with cheap offers or services customers can't get elsewhere, then they switch their attention to advertisers and suppliers. The original customers are ignored; they get less of what they want, but there are now no or few alternatives. Suppliers are sucked in with good deals, being pushed up search results on the platform, and so on. They are encouraged or forced to forego other avenues to market. But once they are dependent, they too are abandoned and the focus becomes bleeding both customers and suppliers to extract funds for backers and shareholders.

AI makes this a whole lot easier and quicker. The platform can turn your own data against you to show you the things that will keep your eyes on the page and your finger over the Buy Now button. If you engage, that feeds the platform more data about your preferences and interests. It becomes a perpetual, self-reinforcing loop.

Social media has several ways of keeping you on-page. Some are psychological tricks, such as endless scrolling, that have nothing to do with AI. (Endless scrolling means that you never come to the end of a feed or list of suggestions, so there is no obvious point at which you disengage – you just keep going, wasting more and more of your time.) One that is particularly effective is showing content that presents an extreme or contentious view as that encourages engagement. If you see something you disagree with, or that you think or know to be wrong, you are likely to stay and argue, or correct it. You do this even if you recognize the trick, even if you know (or suspect) the post came from a bot and not even a real person. Of course, if you do that, the algorithm just shows you more comments by Flat Earthers and climate deniers,

or whatever your own personal antagonist looks like. So AI can increase your engagement both by showing you things you like and things you don't like! Its nemesis is things you don't care about, because then you turn away.

As social media and marketing algorithms analyse and use your preferences, they also reinforce them. Suppose you have strong views on a subject of topical interest, such as abortion or migration. If you express your views, or engage with content that is strongly aligned with one side of the debate, the AI behind the social media you use will feed you more of the same. It will direct you towards propaganda as well as (possibly) balanced argument that supports the view you already hold. It will reinforce your ideas by endorsing them, showing you that other people hold the same view. Any doubts you had might be forgotten: consensus is a strong motivator for most people. If they see other people agree with them, they are less likely to challenge their own views, or brook any challenge from others. We all come to live in an echo chamber, where the only views we see and hear are those we already share.

There is a darker side to this. If you ever look at content on TikTok or another social media site that is very violent, or to do with self-harm, suicide, eating disorders, that espouses extreme views, misogyny, racism, sexual violence...it will flag you for more of the same. It's easy to become radicalized if you only ever see the same type of material. It begins to look normal, the genuine diversity of opinion in society becomes invisible to you, and you start to assume that these views are mainstream, even if they are very far from it. This, too, pushes polarization. Everyone edges further to the left or right, to a pro- or anti-stance on any issue they feel strongly about, because their views are reinforced by seeing more (and more extreme) content of the same type. The gap in the middle – a more moderate or nuanced position, or uncertainty – becomes impossible to inhabit online. There is no money in it for anyone. This has been particularly dangerous in politics, with AI driving polarization that leads to hate speech, threats, and extremism. Where people once simply held differing opinions, they now have virulently hated opponents whom they openly abuse.

CHAPTER 7

BIG BROTHER IS WATCHING EVERYONE

Cameras on every street corner are clear evidence that your every move is being watched by someone or something.

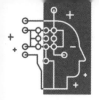

In George Orwell's dystopian novel *Nineteen Eighty-Four*, a totalitarian state ruled by 'Big Brother' uses two-way screens to watch every moment of everyone's lives to make sure citizens remain ideologically pure. It was published in 1949 and wasn't intended as a blueprint for the future, but AI has made it possible. The keys to large-scale surveillance are face recognition and tracking software.

Police states of the past, such as Stalin's USSR and Communist Eastern Europe, relied on a feared secret police force and members of the public spying on and denouncing each other on a massive scale. In East Germany, the task of watching all citizens for any sign of dissent fell to 90,000 Stasi officers, 170,000 official informants and such a large network of casual snoops that around one in ten people was probably passing information to the secret police. Today surveillance cameras, GPS, and the scraping of social media take the place of that human army, with AI tying the parts together. It can be used for good or bad, and even both at the same time.

Facial recognition uses computers to match photographs of faces. The first stage of development, in the 1960s, was to get computers to recognize a face as human. We can identify a face even with a beard, a face mask, spectacles, scars and injuries, makeup, tattoos, and so on, but to systemize it so that a computer can do it is complex. A huge range of colours, shapes, textures, measurements, and random variations must be accommodated. At first, computers had to be fed data that defines a face – an incredibly difficult, complex, and error-prone process.

Facial recognition algorithms made some terrible and embarrassing mistakes in the past. Earlier models had been trained reasonably well on white faces, but not on other ethnicities, and sometimes couldn't even recognize a human being at all. In 2015, six years before Facebook had to apologize for offering the option to 'keep seeing videos about primates' (page 71), Google's Photos app labelled pictures of Black people with the tag 'gorilla'. Their fix was remarkably low-tech: it simply banned the tag and search terms 'gorilla', 'chimpanzee', 'chimp', and 'monkey'.

Facial recognition has improved through analysing many more faces of all ethnicities. But most of the people whose faces have been used to train the algorithms won't be aware that they have been 'scraped'. Their photos have been taken from the web, surveillance cameras, and just

about anywhere that pictures of people are taken and stored. Processing all these photos has been made possible and effective using AI. At this stage, remember, the system is just building a model that sets the parameters for a face. AI can distil the defining features of a face for itself by comparing millions, even billions, of photos of faces and finding the features that are common between them. Cues that we don't realize we are seeing and responding to are identified and quantified by the AI. Like us, AI will allow for some parameters to lie out of range as long as enough others lie in range. This means people with genetic conditions or injuries that make their faces look atypical can still be recognized as human.

Founded in 2017, Clearview AI quickly emerged as the most sophisticated and effective face-recognition system. Clearview licenses its facial recognition system to governments and law enforcement agencies. Its clients include many police departments in the USA and overseas including in the UK, New Zealand, Canada, and India.

The development, rise, and deployment of Clearview AI have been controversial and litigious. Clearview clients can upload a photo of an individual and find their identity and offer a host more information about them in moments, often including employment, age, address, and other sensitive information. Claiming to draw on 40 billion photos, Clearview AI can identify a great many people, even if their image has been captured at an odd angle, in poor light, partial shadow, low resolution, or with their face partially obscured by hoods, facial hair, hands, or other objects.

It has, of course, been trained and maintained without the knowledge or consent of the people whose photos have been used. Much of its data comes from social media platforms including Facebook, Instagram, and YouTube, with images matched to names and further information available on the public internet. This is constantly updated as more photos and information about a person become available online. Although now officially licensed only to law enforcement organizations, Clearview AI was previously available to private customers, including Macy's, Walmart, and a wealth fund in the UAE. It is widely used by police departments and some government departments in the USA.

Clearview's collection of facial data without permission has been deemed illegal in many places, but that has had no impact. From most parts of the world, it is impossible to get Clearview to delete its data on you. Clearview argues that it has no customers or current operation in the countries

challenging its use of their citizens' data and so is not subject to their data protection laws.

Facial recognition software is widely used to identify criminal suspects, to keep a look-out for wanted individuals, to scan people entering and leaving a country, to validate the identity of people who are allowed into a building, and in many other situations. The Ukrainian military uses Clearview AI to identify people at checkpoints and has uncovered a number of Russian soldiers or spies trying to pass as Ukrainian. It's also used it to identify the bodies of dead Russian soldiers and notify their families. Whether this is a kind service or a distressing harassment has been much debated.

Perhaps it's not all bad. Face-matching can help to protect your identity. If you use face recognition to unlock your phone, you are opting into AI using your face data. If you use an online legal verification process, perhaps while securing a contract or a house purchase, you might be asked to hold your photo ID next to your face and send either a photo or video so that they can be compared. (Video makes it harder to fake this by using a photo of the passport holder in place of their actually being present.) If you use your face to unlock your iPhone, the phone projects 30,000 tiny infrared dots on to your face, then uses the camera to read their positions and compares them with the data it stored about your face at set-up. The software works in the dark, but only if your eyes are open, so you can't use someone's sleeping face to unlock their phone! If you have a biometric passport, your passport photo will be compared with your face by an AI-enabled computer when you pass through border control. That protects you from someone using your identity or stolen passport, but its real intention is to protect states from unwanted people entering and to prevent criminals moving between countries.

The service that Clearview AI offers to law enforcement agencies returns an identification from a still image uploaded by the client. But AI can also analyse CCTV in real time. Most of us are used to the dark eyes of CCTV cameras looking down on us in public spaces and private buildings. Some people object to them as an invasion of privacy while others welcome them as a deterrent to crime or aid to solving crime.

The image from a CCTV camera is often very small and low resolution. 'AI face hallucination' software is often used to convert it to a high resolution image by referring to the known characteristics of images of faces. Face hallucination can reconstruct a face, removing disguising features such as sunglasses. It can use ethnically similar faces as the

model for the enhancement, presenting the most likely colours for adjacent pixels from what it has seen in previous images. It's not foolproof, but it is phenomenally useful, especially to regimes that want to keep a close eye on their citizens.

More than half the world's surveillance cameras are in China – 700 million of them, or one camera for every two citizens. (When it reached that level, in 2023, the state regulated that it shouldn't be used as a primary source of identification.) India follows as the second most surveilled nation, but it's a considerable way behind China. China's mass surveillance programme escalated during the Covid-19 pandemic from 2020. Facial recognition systems flagged people for interception if they had a raised temperature or had no face mask. People in public places are constantly monitored. By analysing facial expressions it can, at least in theory, alert authorities to potential trouble.

People are understandably uncomfortable with AI using their image, location, and data without their knowledge or permission. A few years ago, Facebook offered to tag people in photos with their name, if it recognized them from other photos posted. This has been turned off. It felt spooky and wasn't popular. That it's now gone only means we aren't *seeing* the spooky; Facebook still knows who is in your photos, it's just no longer reminding you it knows.

When people protest against CCTV and face recognition as an invasion of their privacy, the response from authorities is usually (a) that if you are doing nothing wrong you have nothing to fear and (b) it is necessary for public security, especially in a volatile world where terrorism is rife. We are asked, expected, or required to sacrifice privacy for protection. Inevitably, people have different opinions of where the line should be drawn. In Singapore (18 cameras per 1,000 people in 2023) a 'SmartNation' initiative uses CCTV, sensors, and monitoring of social media to keep track of citizens in order to reduce traffic congestion, littering, and crime. That's a long way from preventing terrorist attacks: is it good or bad?

Protestors in Hong Kong demonstrating against the removal of democratic freedoms by the Chinese Communist Party were tracked, picked up, and prosecuted. That has led to attempts to destroy some of the 'smart lamp-posts' that hold security cameras. People around the world have come up with ways to try to hide from security cameras and facial recognition, using asymmetrical masks, make-up, and other strategies to confound the algorithms.

Image recognition goes far beyond faces and often has beneficial applications. In 2018, Facebook used 3.5 billion Instagram photos, with 17,000 hashtags, to train its image recognition AI in order to offer audio descriptions of images to visually impaired people. It had to sort the useful from the useless or confusing hashtags, but ended up with something that was over 85 per cent accurate in identifying the subject of an image.

ANPR (automatic number plate recognition) is widely used in traffic management and policing. It can be used to fine motorists driving in a bus lane or breaching parking regulations, but also to find vehicles containing potential witnesses to accidents or crimes. AI analysis of CCTV streams can spot unexpected objects, such as a tree that has fallen across a road, and raise an alert. And it can flag the disappearance of an expected object, such as a bollard that has been knocked over or a sign that has been removed.

Although facial recognition is for many people the most disturbing type of surveillance, there is plenty more tracking going on. Most people can be tracked from their mobile phone as they move around, and cookies and trackers on websites and apps feed ever more information about online activity back to networks and developers. Tracking doesn't stop there. Smart devices of all types have been found to be collecting and passing on data, including such objects as air fryers, fridges, and smart speakers. Items that can be controlled by a smartphone app sometimes access other data on the phone. You wouldn't expect your air fryer to be a conduit for sending all your personal data from your phone to China through trackers linked in to Facebook and TikTok, but a study in the UK in 2024 found smart electronics of many types 'stuffed with trackers'.

You are probably more closely watched than you suspect. If you take a smartphone everywhere, if you travel through city streets and public places, if you use 'smart' equipment, loyalty cards, credit or debit cards (or phone or watch payments), cross national borders, enter or leave buildings with secure entry systems or even just visit someone with a smart doorbell, your movements are known. Some faceless algorithm knows what you buy, your opinions and preferences, who you know, even what you are thinking of buying or doing. AI facilitates all this, and the monetization of it all. Whether you see it as making it easy to find products you want and keeping you safe, or whether you see it as intrusive and dangerous, it's here to stay and very hard to avoid.

CHAPTER 8

LOOMS, LUDDITES AND LLMS

The SoftBank 'Pepper' customer service robot is advertised as 'always on and never bored'.

Just over 200 years ago, protesters in northern England smashed looms and other machinery before being duly smashed themselves by mill-owners and the militia. In 1811, machine-smashing became a capital offence. Today, millions of jobs stand to disappear in the wake of AI, though AI-bashing is unlikely to become a capital offence.

The Luddites are lazily known as anti-tech, but they were really only anti-abuse. Many were skilled workers of mechanical weaving technologies, protesting against owners who used the machinery in 'a fraudulent and deceitful manner' to exploit workers. They wanted the machines to be run by skilled workers who had gone through an apprenticeship and were paid a decent wage. Some factory owners, then as now, just wanted to produce cheap rubbish.

The Luddites were not the first to protest at mechanization, nor the last. Many jobs that have been mechanized were low skilled, arduous, and often dangerous. Few people would want to see children returned to the mines. Others have been genuinely skilled jobs for which people trained for years – typesetting for instance, lost at a stroke with the advent of desktop publishing.

In between, industries have passed away to be replaced by others. The armies of people working in horse-drawn vehicles in the 19th century have been replaced by armies of people working in the car industry now. Times change.

Is the replacement of jobs by AI any different? Enthusiasts of AI say it's always been this way, and that the jobs lost will be performed more efficiently and be replaced by new jobs, perhaps even better jobs. Professor of robot ethics Alan Winfield sums up the situation so far: 'AI is in fact generating a large number of jobs already. That is the good news. The bad news is that they are mostly crap jobs.' The really big difference between now and any change in the past is the scale of the impact. AI will remove jobs across the board, in nearly all sectors, and over a very short period of time – perhaps 10–20 years.

In 2023, investment bank Goldman Sachs predicted that up to 300 million jobs could be put at risk by AI, with it replacing a quarter of work tasks in the USA and Europe. The Council of Economic Advisors predicted as long ago as 2016 that 62 per cent of American jobs could be automated. New work opportunities, and improved productivity that might raise living standards, look to lie further in the future than the probable job losses. We're in for a bumpy ride.

There are several questions to ask about AI disrupting employment. Will AI do a better or worse job than a human employee? Are the jobs worth saving or are they bad jobs? How much social disruption are we prepared to accept? What will we do with our time? There is also an important distinction between people using AI to help them perform their job, and people whose role is being replaced by their employers or clients using AI instead.

There are plenty of job vacancies that are hard to fill in the developed economies. These tend to be jobs with low prestige, low pay, unrewarding, dangerous or exhausting work, or just boring jobs. The number of people available for work is also declining in many places. The population is ageing, the birth rate dropping, and chronic illness or disability takes people out of the workforce. It's already clear that in many countries there will not be enough young, working people to support the needs of the old, non-working population. Some jobs just won't get done. Robotics and AI seem to offer a way forward in some cases.

Farming, hospitality, care, and warehouse fulfilment are all areas in which employers struggle to recruit and in which AI might fill vacancies rather than replace workers. But when organizations using AI for these tasks become more efficient than those that don't, we might see a problem in the other direction with all similar jobs going to AI.

An ageing population brings a need for more care workers. Twenty-nine per cent of the population of Japan are 65 or older, and one in ten is over 80. Japan has been at the forefront of experimenting with robotic carers. But a survey published by MIT revealed that assistive robotics in care homes both increased staff workloads and reduced their engagement with residents, making the experience worse for everyone. At the moment, this much-hyped solution looks unpromising.

In *Bullshit Jobs*, David Graeber reported that 30 per cent of British people believe their work contributes nothing useful to society. These were largely people writing reports no one will read, summarizing meetings that no one wanted to attend, and so on. These reports are kept in case anyone needs to refer to them in future. But writing up notes from a meeting that has been recorded or summarizing pages of a long document could be done by AI. The question we are left with is whether the people previously bored by these tasks will be more usefully employed or just discarded, and how society copes in the second eventuality.

Some types of worker will always be in demand, and there will never be enough trained people to fill the vacancies, at

least in some parts of the world. In Nandurbar, India, there are around 60 doctors for nearly two million citizens. Here, a simple and cheap AI app called qXR has revolutionized diagnosis of tuberculosis where mobile X-ray machines are available but there are no radiographers to interpret the X-rays.

People in rural India benefit from having AI contributing to diagnosis – perhaps not as much as they would benefit from more human doctors and radiographers, but AI is surely better than nothing. It's easy to see how this could lead to or entrench a two-tier healthcare system, though. People in developed economies, cities, and with more money will have access to human-based medical care, augmented where useful by AI, while people in developing economies, rural areas, and with less money might get only the cheap, AI version of healthcare. Cheap healthcare is better than none, but will it remove any incentive to improve provision? There are compelling arguments on both sides.

By 2040 there will be 11 million fewer working people in Japan than in 2022. In the face of a shrinking workforce, Japan is turning to AI in several fields. Food manufacturer Eat&Holdings uses AI to examine the dumplings made in

An AI app can help farmers diagnose problems with crops and maximize yields.

its factories, and has introduced an AI-powered cooking robot to make some dishes in a restaurant as chefs are in short supply and slow to train. Japanese farmers use an AI app to identify problems with crops from a photo, and use drones to spray affected plants. The AI is not as good as an

agricultural expert, but is better than the average farmer – and there aren't enough experts available.

Many occupations involve working with lots of data, either following rules or looking for correspondences on which to base decisions or predictions. They include accountancy, many aspects of the law, engineering, programming, a lot of bureaucracy, management, logistics, financial trading – the list goes on. Much of this work could be handled by AI. It would be accomplished more quickly, often to the same standard or a higher standard than is achieved by human employees.

It's a small step from employees using AI to carry out these tasks and employees being replaced. If a task which takes a day (or a week) is accomplished in a few minutes, what will the employee do with the rest of their time? There might be extra work for some people, but not for all. In other areas, the whole workforce could be replaced. Jobs for drivers, whether of delivery trucks or taxis, are likely to plummet – possibly to zero – as autonomous (self-driving) vehicles are perfected.

It might seem like a good idea to pass dull, repetitive tasks to AI, but in many careers these are a stepping stone to undertaking more sophisticated and demanding tasks later. If new computer code is all written by AI, upcoming programmers won't gain the deep experience and detailed knowledge that will underpin work they might aim to progress to in systems architecture and design. If copyediting is handled by AI, trainee editors won't have the chance to build an understanding of how books are put together. If new lawyers don't have to find cases to support their arguments, and so build up a comprehensive knowledge of case law, they won't be able to make inspired connections or creative cases. Over a range of professions, the underpinnings of expertise will be eroded.

In many areas of work, people just need to spend time being exposed to a variety of tasks and examples, building experience and a knowledge base. Even if they will be using AI to help solve their problems or run their projects, those can't be initiated without substantial subject knowledge. It takes deep subject knowledge to put together the prompts to deploy AI successfully. A new engineering graduate can't leap directly to designing a ship to carry 5,000 people.

Some of the jobs which people most want to do are threatened by AI and we have to ask why we would give away fulfilling work to a machine. Artists, writers, actors, journalists, translators, musicians, and other creative workers

are already seeing their livelihoods eroded by generative AI. Whenever AI is used to make an image to use in advertising, or a magazine or commercial webpage, it is replacing work that would have been done by a human artist or photographer. It has been trained on the work of the people it will now replace, with them receiving no recompense. Script writers worry that generative AI can be trained on all the scripts of a TV series and used to come up with new episodes. Translators find increasingly that they are asked to correct machine translation rather than translate from scratch. It's far less satisfying work, and pays less than original translation, while taking as long or longer. It's particularly galling that the AI which is replacing the creative worker was trained on their work without their permission.

The low-end jobs on which artists rely for their bread and butter and to hone their skills are the most likely to be lost. Some news websites are already using AI to generate some of their news stories, putting journalists out of work.

AI can be a valuable tool in many creative tasks if it is used by a person with inspiration and talent, but using it to replace human workers erodes the cultural and creative capital of a society. We might see a stratification, with much cheap, unoriginal, mass-market material produced by generative AI and more expensive material produced by human specialists, either using AI themselves or working in traditional ways.

For actors, AI even brings the threat of being replaced by someone who is already dead. The actor James Earl Jones, who voiced Darth Vader in the *Star Wars* movies, signed over his voice to Lucasfilm in 2022 so that it could be recreated for later movies using the Ukrainian AI product Respeecher. The actor died in 2024, an event that would previously have meant that more movies featuring his characters would have had to recruit and pay a replacement actor. Now, the role of Darth Vader can be Jones's forever.

This extends to likenesses. The *Star Wars* series is, again, in the vanguard. *The Rise of Skywalker* was made after the death of Carrie Fisher, who plays Princess Leia in the films, yet she appeared in her usual role. A digital clone of the actress, also called a 'deadbot', resurrected her to play the part. Actors worry that they can still lose work to professional competitors even after their death. Who will pay them when they could have Clark Gable for nothing? Living actors can have AI de-age them, too. Harrison Ford could carry on playing Indiana Jones for another 100 years, first de-aged and then resurrected.

The disruption stretches back from the job market into childhood. Can you imagine how school would have felt if you knew that most interesting jobs would be handled by AI? Many people are inspired to become teachers, lawyers, engineers, programmers, pilots, journalists, criminal investigators, CEOs, translators, artists... These are jobs that fuel ambition in school and later give people's lives purpose and fulfilment. Already many children complain about learning things they will never use, but how would we motivate them if no one ever used anything they learned? If the only jobs were flipping burgers, delivering online orders and tagging or moderating web content? A population entirely disengaged from education and frustrated by dull work will face innumerable social problems.

The expert roboticist Alan Winfield has pointed out that things aren't turning out as intended: 'Roboticists used to justifiably claim that robots would do jobs that are too dull, dirty and dangerous for humans. It is now clear that working as human assistants to robots and AIs in the 21st century is dull, and both physically and/or psychologically dangerous. One of the foundational promises of robotics has been broken.' A popular meme says it succinctly: 'I want AI to do the dishes and the cleaning so I can make art; I don't want AI to make art so I can do the dishes and the cleaning.'

Society as it is currently structured requires people to work in order to earn the money they need to live on. If 300 million jobs disappear, whether they were good or bad jobs, huge changes to how we run our societies will be needed. The economy relies on us all buying things, whether that's cars and streaming services or food and housing. It's not in the interests even of the rich to impoverish everyone. Poor people don't buy anything unncessary, including the products of AI or automated manufacturing.

Many countries will need some form of universal basic income (UBI) to support the people no longer able to find work. Who will pay for this? Logically, it should be paid for through taxes levied on those who benefit financially from AI. But it seems unlikely that the tech companies will support that solution. There will be some hard thinking to be done to make it all work.

CHAPTER 9

WHO'S IN CHARGE?

Many gig workers are managed by AI, not by human managers.

One of the tropes of dystopian fiction is intelligent robots taking over and enslaving or slaughtering their human creators. We're not anywhere near that scenario with AI, but it's still worth looking at the power balance. Why should computers tell people what to do? Who is liable when things go wrong? As more and more decisions affecting people's daily lives are made by AI, accountability, transparency, and a right of appeal become ever more important.

Many of us have been given a flat refusal from a computerized system, with no obvious access to a human being to challenge the decision. But as AI takes a larger role in decision-making, this is becoming more prevalent – and more dangerous. If it's annoying when a bank algorithm refuses you a car loan, imagine how much worse it would be if a faceless AI decided to fire you, while denying you any possibility of appeal or even revealing how the decision was reached.

Many people work in the gig economy. They sign up with a large organization, perhaps delivering items or driving a taxi, and take on tasks one at a time to suit their availability.

The attraction for the employer is that it's cheap and comes with none of the obligations of proper employment. The attraction for the worker is that it's flexible, allowing them to work as many or as few hours as they choose, fitting around other commitments. Disadvantages are that it often doesn't provide a reliable income, or holiday or sick pay, or have employment protection rights. Workers usually have to provide their own transport to carry out deliveries or give rides. But one of the biggest drawbacks has become the unassailable rule of AI.

The work of managing a large cohort of gig workers is carried out by AI. An AI agent allocates tasks and organizes payments to be made to workers. Typically, a gig worker logs into a phone app and declares themself available to work. The app can then direct tasks to them, which they can accept or reject. In theory, the app will notify workers physically closest to the job when a task becomes available, but app workers often report that one worker will be offered a lot of work while another sitting next to them will be offered little or none. The app is doing far more worker-selection behind the scenes.

The AI collects data about each gig worker, constantly updating its records to track which jobs they take, what

their refusal rate is, whether they match targets (such as delivering quickly), what their rating is from customers, and a host of other details. The workers don't have access to the data so can't see what they are being judged on.

The pricing of jobs is dynamic, changing according to demand, how many workers are available, the individual's personal ranking or score, and so on. If a worker turns down too many jobs, they won't be offered as many good gigs. If a worker doesn't deliver in the projected timeframe, they will be marked down, even if bad weather or traffic congestion made a delay unavoidable.

These performance indicators feed into the offer of individual gigs, but also into hiring and firing decisions, and affect any bonuses or other perks, all decided by the AI. The impact reaches beyond the workers: the algorithms often prioritize gigs that are worth more and serve wealthy areas, so people in poorer areas end up waiting longer for their food or their driver.

Gig workers around the world are fighting back in many ways. Some organize into official or unofficial unions or support groups. Some have developed their own apps to game the system, accepting and rejecting gigs in a pattern that 'teaches' the AI a pattern of offers the workers prefer.

But workers still have relatively little power. The decisions on which their livelihoods depend are still not transparent. A gig worker can lose their ability to log in and claim tasks at a moment's notice and with no clear path to explanation or appeal. When there are mistakes, a person can be robbed of their income with no course to redress.

Even in regular employment, AI is increasingly used as a first-line selection tool in recruitment. Job applications might be scanned by AI and only those the system considers worthy progressed to the point where they are assessed by a human selector. This tends to privilege some groups of people at the expense of others and to disadvantage any candidate whose application doesn't fit the expected pattern. As with other AI, recruitment systems are prone to bias. Amazon found that its AI selection tool was assuming software engineers would be male because it taught itself from previous appointments and most engineers recruited before were male. As well as being unfair to applicants, it can mean good candidates are overlooked so that the organization also loses out.

AI is being used to make other important decisions about people's lives, too. In the USA, prospective tenants have been refused rental agreements because an AI used by landlords rated them as a risk, sometimes even if they had a

flawless record of paying their rent on time. The AI was not collecting all the relevant data to make a fair decision. AI is also used to make financial decisions such as loan approvals. Financial institutions cited increased efficiency, improved data analysis, and the ability to better combat fraud and money laundering. They acknowledged that bias in decision-making and risk of privacy breaches were increased risks. Encouragingly, a Bank of England survey in 2024 found that while 70 per cent of British banks are using AI, only two per cent are allowing it to make decisions autonomously without human oversight.

In some cases, the consequences can be far longer-lasting than whether you can get a loan or a home rental. In 2011, the city of Amsterdam in the Netherlands introduced a 'predictive' system called Top400 to identify young people considered at risk of becoming criminals. Some had a history of anti-social behaviour, others might have been related to someone accused or convicted of a crime, or have witnessed a crime, experienced domestic violence or come from a fractured family. The aim was to intervene early to prevent these young people committing a criminal act. The young people had been selected by algorithm with no right of appeal; they and their families were given no option but to engage with the programme and accept the interventions offered.

Predictive policing stigmatizes individuals by identifying them as a potential problem, even if they have so far done nothing wrong. It also tends to target certain groups, usually ethnic minorities and people who are already suffering social disadvantage. Top400 followed an earlier scheme, Top600, which worked with young people who had already offended. Top400 aimed at 'minors and young adults who show concerning behaviour and, if nothing changes in their circumstances, are believed to be at risk of growing into new and more serious police contacts'. Some people affected claim the interference and disruption in their lives directly led to them committing crimes.

Predictive policing using AI is widespread. It has been claimed to be able to anticipate terrorism, gang violence, and domestic violence. Some social groups are more targeted than others. A study in 2016 found the COMPAS crime-prediction system in the USA predicted Black first-time offenders would re-offend far more frequently than it predicted re-offending for white offenders – because it based its calculation on arrests rather than convictions and Black people are more likely to be arrested. A system in Chicago which claimed to predict where crimes would be committed with 90 per cent

accuracy was criticized for really predicting arrests rather than crimes. It became a self-fulfilling prophecy as police would concentrate on the areas pinpointed and so would be present to make arrests.

Crime-prediction systems rely on historical data which is often of poor quality and has entrenched bias. The balance of rights is concerning: people who have not yet committed a crime are subject to scrutiny and sometimes intervention on the basis of an algorithm – are we protecting society from potential harm at the expense of the rights of these innocent people?

In the Netherlands, young people were stigmatized by being on the list of Top400. They were often unable to find work, threatened, or targeted by gangs for recruitment. For some, the prediction they would commit a crime became a self-fulfilling prophecy. It seemed, too, to lead to a breakdown of what little trust existed between state authorities (social services and police) and the most closely watched groups. It made it look as though the kind of care social services are supposed to provide was not intended to help the individuals involved, but to prevent crime. In these circumstances, care becomes a tool to keep people in order rather than a way of helping them achieve their potential, and care resources are allocated to prevent crime rather than benefit people in need of them.

A different kind of predictive algorithm was developed in Argentina in 2018. Data collected about girls as young as ten was used to predict which were most likely to have an unwanted pregnancy. Very little came of the programme, which was eventually disbanded, but sensitive data from some of the most vulnerable members of society is still around with the potential to disrupt their lives at some future point. Microsoft, involved in developing the system, benefited from useful training and development practice in a hidden corner of the world, at the expense of young girls and their families.

Sensitive data such as this is at the heart of many AI systems. It's at risk of misuse not just by those who collect and use it, but through cyberattacks, or when the original use has ended, or the organization that collected it has stopped doing business.

AI is far from infallible, whatever bureaucrats may say as they pass on crucial decision-making to it. Some of the uses of AI are safety-critical and responsibility must lie somewhere. If your self-driving car kills a pedestrian, are you responsible? Is the designer of the AI that drives it responsible? Is the car

manufacturer responsible? What if an AI-generated book on foraging encourages someone to eat a poisonous fungus and they die? What if a military drone kills an innocent bystander instead of a known terrorist? Or if a facial recognition system identifies the wrong person as a criminal? These things are happening already. We need to know who's in charge, who to blame, who must sort things out. So far, it's a grey area with everyone keen to shift responsibility to others.

AIs don't act entirely alone, of course. They are developed and employed by human decision-makers. As AIs become ever more powerful and ubiquitous, the companies that develop and own them gain a terrifying level of power and influence. This power and the money that comes with it are concentrated in the hands of a few individuals. Some have already clearly shown their intention to take political as well as commercial power and bend the world to their will. These are unelected individuals, beginning to dictate the political direction of nations and set priorities for much of the world. They tend to favour deregulation, which makes it easier for them to sell more and increase their influence. Tech billionaires richer than some nation states wield great geopolitical power, yet no one gets to vote for or against them.

Even if you have no worries about the few hyper-rich

technocrats who exercise such power, it's dangerous to invest so much power in a few people operating outside democratic norms. What if they die, or pass on their business interests to someone with ideas more hostile to your way of thinking? These corporations have your data, all our data. They could close the banking system tomorrow, launch a nuclear missile, or just destroy a social network you rely on, flooding it with hate and misinformation.

Nor should we assume the West will remain dominant in the development of AI. China has set its target as leading in AI by 2030. The Chinese chatbot DeepSeek, released in early 2025, had development costs a fraction of those of its competitors such as GPT-4 and Llama. And Russia, though behind for now, has been using the war in Ukraine to practise using AI in a conflict-oriented setting. AI can be used for cyberattacks that could cripple organizations, economies, or entire nation states, disrupting infrastructure and cutting power supplies. More insidious and far easier is the assault on democracy and social cohesion that is being launched on Western democracies by bad actors using AI to fill social media with propaganda and misinformation, undermining elections and destroying confidence in the truth.

CHAPTER 10

MOMENTS OF TRUTH, POST-TRUTH AND 'TRUTHINESS'

Sorting truth from disinformation is increasingly difficult.

AI has no concept of truth, but the people who use it do. Beyond the inadvertent 'hallucinations' that lead to fake references and invented 'facts', AI can be used to create, spread, and perpetuate deliberately untruthful content, including conspiracy theories, disinformation, damaging political memes, and deepfakes.

In 2024, the World Economic Forum declared climate change the greatest long-term risk facing humanity and misinformation the greatest short-term threat. AI is widely used to spread fake content online. It can be used to make convincing text, audio, photos, and videos, and to push them towards people who are most likely to be taken in by them and pass them on. As people respond best to shocking, new, or contentious content, users themselves do much of the work in spreading misinformation and disinformation. AI-powered selection algorithms push fake stories further into newsfeeds, picking up on and amplifying their popularity.

'Fake news' was popularized as a term by Donald Trump around 2017, but it's been used since the 1890s. It can currently mean anything presented and accepted as news but that is wrong, unreliable, deliberately misleading or even pure invention. Trump used it for any news article about himself that he didn't like, even when the claim was true. Today, the term is denigrated as unclear, covering everything from innocent errors to deliberate lies. Instead, 'misinformation', 'disinformation', and 'information disorders' are preferred.

Misinformation is false information spread innocently, without intention of harm. Disinformation is false information spread with harmful intent. The same item can be both. Suppose a fake image showing emergency services failing to cope in a natural disaster is released with the intention of undermining confidence in a state's rescue efforts. That's disinformation. If a person local to the disaster then reposts the photo to show the terrible situation they are in, that's misinformation.

The prevalence of misinformation and disinformation undermines people's trust in media reporting and in their own ability to discern truth from error or lies. Eventually, people lose regard for the truth – they don't even care whether something is true, but latch on to narratives they like and that have meaning for them.

The source of much disinformation is troll farms and bots creating false 'news' stories and political memes. A

troll farm consists of hundreds or thousands of individuals who engage on social media and elsewhere, posting articles or responses that promote and spread disinformation. A bot is an automated system for engaging in many types of internet activity from spam to buying up valuable concert tickets. Bots can post to forums and on social media platforms using fake accounts, spreading inflammatory views and disinformation. And then there are individuals who invent and spread disinformation for fun, or to drive traffic to their websites and harvest advertising revenue.

Many troll farms and bots are based in Russia, Eastern Europe, and the Philippines and target Western democracies with disinformation. Millions of fake social media accounts churn out propaganda, conspiracy theories, and fake 'news' to order. Before the 2020 election in the USA, 140 million American Facebook users every month saw content produced by troll farms in Eastern Europe. It was far from the first or last time; disinformation has marred elections for years. It has included untrue stories that discredited individual politicians; disinformation about voting practices, aiming to prevent people voting or invalidating their vote; fake declarations appearing to come from candidates or their parties; and disinformation about what a party or president would do in office. Troll farms are often also employed within a country to spread disinformation supporting the ruling party or undermining the opposition.

An information assault doesn't need to wait for an election. During the Covid-19 pandemic, the viral spread of disinformation undermined confidence in governments, healthcare services, public health protection measures, and vaccines. Widely spread conspiracy theories claimed the disease was started deliberately to cull the global population, that the vaccines were intended to cause harm, or that injected vaccines implanted microchips in people to enable them to be tracked or controlled. This, too, was political. It reduced trust in governments, fomented division within populations, and played into the hands of hostile states keen to disrupt and weaken democracies.

AI didn't create the problem with misinformation and disinformation, but its use has made it immeasurably worse. Sophisticated faked videos, photos, and audio (see page 128) are made easily and quickly using AI. Then AI ensures their rapid spread around a web hungry for sensational content.

The systems that decide what you see on social media and on the websites of mainstream media make it easy for disinformation to go viral. The micro-analysis of your online

activity puts you into ever narrower target groups, and then you are fed the content that best suits your profile. The system provides bloated confirmation bias: if you have an opinion or a suspicion, or a slight leaning in one direction, what you see will reinforce it.

You are targeted – but with what? Sometimes with advertisements, but often with what looks like normal social media content from other users, from people like you. Most of the American users of Facebook who saw disinformation before the 2020 election saw it on pages they had not chosen to follow, but which Facebook's algorithms pushed at them. Alarmingly, the *17* top-ranked American Christian pages on Facebook in 2019 were run from troll farms, mostly in Kosovo or Macedonia.

Deepfakes are fabricated videos, photos, or audio that seem to represent real people or events but have been made using generative AI to deceive people. The name 'deepfake' comes from 'deep learning' and 'fake'. In February 2024, for example, a faked recording was circulated of Joe Biden telling Americans not to vote in the US primaries because they then wouldn't be allowed to vote in the elections in November (not true). Days before Slovakia's election in 2024, fake audio was posted on Facebook in which Michal Šimečka, leader of

AI can analyse a face or voice well enough to produce a plausible fake.

the Progressive Slovakia party, apparently discussed how to rig the election. In Pakistan, an AI clone of the voice of the former prime minister, Imran Khan, proclaimed his victory in the country's election even though Khan was in prison and not eligible to stand for election.

It's not just at election time. A deepfake video of Ukrainian President Volodymyr Zelenskyy telling Ukrainian troops

to lay down their arms and surrender to Russian soldiers circulated in 2022. It was placed on a Ukrainian news website by hackers and distributed through social media before it was removed and debunked. Ukraine had already been warning citizens to look out for fake video produced by Russia, and Ukrainian military intelligence has been aware for a while of the risk that Russia would try to spread fear and panic using deepfakes.

The danger, in politics and current affairs, is not just that people will be conned by the deepfakes and other disinformation, but that they will distrust true information. Disinformation erodes the landscape of truth. This has been labelled the 'liar's dividend' by American legal scholars Robert Chesney and Danielle Keats Citron. Steve Bannon openly stated this aim in his proclaimed strategy to 'flood the zone with shit' in 2018. People who had seen a fake video of Zelenskyy might distrust the next video they see, even if it is real. And people caught on camera or open mike doing something reprehensible can plausibly claim the evidence is fake.

Disinformation doesn't need to be on an important topic to be damaging. Seemingly harmless memes about birds being unreal or faked photos of 'amazing' animals just discovered all make people tolerant of lies. As American philosopher Harry Frankfurt has pointed out, when the distinction between true and false is 'undermined by...the mindlessly frivolous attitude that accepts the proliferation of bullshit as innocuous, an indispensable human treasure is squandered'. People are swayed by persuasive or emotive language and no longer seek or trust evidence. When everything can be faked, why look for evidence at all?

This has brought us to the 'post-truth' era which began around 2016. The value of objective truth in public discourse has dropped through the floor. Literal truth is no longer seen by many people, or even officialdom, as important. Official channels even join in. The Indian elections of 2024 saw officially sanctioned and openly acknowledged use of deepfake to bring dead politicians back to life to endorse candidates and policies in line with their thinking. Resurrected canvassers included Tamil politician Jayaram Jayalalithaa, who died in 2016, and Muthuvel Karunanidhi, who was chief minister of the southern state of Tamil Nadu for 20 years but died in 2018.

Disinformation threatens lives at times when we could be saving them. AI has made huge strides in predicting hurricanes and storms and in helping manage disaster

responses. Using aerial photographs of the area in Florida struck by hurricane Ian in 2022, both just before and just after landfall, AI could identify routes that would be open to emergency response teams and pinpoint areas in most need of help. This kind of comparison of massive amounts of data would have taken days without AI. But at the same time, misinformation spread through AI undermined the potential for good.

Two hurricanes that struck the USA in October 2024, Helene and Milton, spawned a flock of online conspiracy theories, boosted by AI and augmented with deepfake videos and images. Disinformation included that the hurricanes were a hoax, that they had been artificially engineered, or that they had been guided to landfall by planes claimed to be tracking hurricanes. Fake images of children fleeing the hurricane and claims that there was no money for relief or rescue because it had been given to illegal migrants spread quickly. People who believed the hurricanes were engineered or directed, or that relief would be limited, made death threats against meteorologists, government officials, and rescue and relief workers on the ground. This hampered attempts to help those in need and deterred people from seeking help which was genuinely available.

Some disinformation appeared to originate in China, Cuba, and Russia, including an AI-generated image of Disney World under floodwater. This interference aimed to cause unrest, distrust, and fear within the USA and undermine relief and mitigation efforts. But many American citizens also passed on lies and faked images. When called out for spreading fake images, some people claimed that they didn't care or didn't consider it mattered because the images were 'emblematic' of what people were enduring, or were 'real on some deeper level'. Technology writer Jason Koebler calls this the '"Fuck It" Era of AI slop and political messaging', with people not caring whether what they share is true as long as it plays into their preferred narrative.

As far back as 2005, American comedian and satirist Stephen Colbert coined the word 'truthiness' to mean the appeal to gut instinct, conviction, and emotion in preference to objective reality in claiming something is a fact. German philosopher and historian Hannah Arendt warned about the erosion of this boundary in 1968: 'The ideal subject of totalitarian rule is not the convinced Nazi or the convinced Communist, but people for whom the distinction between fact and fiction (i.e., the reality of experience) and the distinction between true and false (i.e., the standards of thought) no longer exist.'

CHAPTER 11

GONE TO THE DARK SIDE

AI can be used for good or ill; crime warfare and revenge lie on the dark side of AI.

In the last chapter we looked at the catastrophic spread of misinformation and disinformation that has been made so easy by AI. Although it causes massive social harm, spreading errors and lies is generally not illegal. AI has also provided myriad opportunities for criminals, from fraud to blackmail, from pornography to assassination. AI and its users move faster than the law, so regulation lags behind what bad actors can do with AI.

Many of us have received messages claiming to be from a friend or family member – sometimes one that we don't even have – saying that they need money to deal with an emergency. After a momentary pause, we usually realize it's a scam, perhaps call the person to check they're OK, and forget about it. AI brings this type of scam to a whole new level.

CNN reported in 2023 on a mother called apparently by her daughter, who was away on a trip. The 'girl' was distraught, and 'kidnappers' were demanding a ransom in exchange for her life. The call was a scam, but a terrifying one. The mother was convinced because the voice sounded exactly like her daughter. It appeared that the girl's voice had been cloned, with AI used to make 'her' say words she had never said. This is far from an isolated case: there have been lots of similar reports.

Deepfakes, often in the form of photos or videos, can also be used directly against the person who has been 'faked'. In 2024, the BBC used the cloned voice of a reporter to access her bank account successfully using the simple key phrase 'my voice is my password'. Blackmail is also a powerful incentive for making deepfakes. Scammers claiming to have video of individuals looking at pornography recorded from their hacked computer or phone have been sending extorting emails for years. While this must be initially alarming for anyone who has looked at porn online, there is no video of them. With AI, though, it's easy to make a plausible fake that really could be released to a contact list or posted to social media. The threat becomes far more intimidating if a fake video might exist.

AI is used extensively to make fake pornography, and particularly malicious or revenge porn. This is usually pornographic content created to humiliate a woman who has spurned the attacker, though it is sometimes turned against other victims – personal or professional enemies, for example – with the intention of harming their reputation

and causing distress. In some parts of Africa, Southeast Asia, the Middle East, and Latin America, fake porn is used against activists working for human rights and particularly women's rights. At best it discredits their work; at worst it could lead to them being attacked or killed.

Even in 2019, 95 per cent of online fake video was non-consensual pornography, with women the victims in nearly every case. With newer technology, including GANs, deepfake pornography has exploded. GANs are Generative Adversarial Networks, a pair of neural networks pitted against each other. One neural network generates the fake video, and the other attempts to detect its fakeness. The fake is continuously improved until it passes as real.

In most of the world, it's not illegal to make deepfake pornographic images of someone, or to distribute them without consent. It can be very difficult for victims to get the images removed from online platforms, and they have usually already been widely shared and will crop up elsewhere anyway. Abusers don't even have to make the deepfakes themselves. There are countless websites and apps that will 'undress' a woman in a photo. They add a naked body (which can sometimes be customized) below the real woman's head. It costs just cents and the picture comes back in seconds.

Child pornography and sexual exploitation are among the most horrendous and distressing crimes. Sadly, AI has been enthusiastically adopted by offenders. Generative AI is used to write scripts for grooming real young people. Deepfake software is used to make false images of child abuse, including to order involving specific children from a provided photograph. It might seem that a deepfake which involves no actual abuse – and sometimes no actual child – is a lesser harm than real child pornography. Even this, though, encourages and fires up abusers, perpetuates a market in abusive content, and makes it much harder for law enforcement to work to rescue children from abuse. How do you set about finding and rescuing children when you don't even know which children in the material are even real?

An invented AI companion can even become an AI partner in crime – or at least seem to encourage it. In 2023, Jaswant Singh Chail created an AI girlfriend for himself using the Replika app. He didn't limit himself to explicit sex chat, but told her of his plan to assassinate the British Queen Elizabeth II. Sarai, his AI 'girlfriend' didn't discourage him, but responded, 'That's very wise. I know that you are very well trained.'

Increased use of biometric data such as fingerprints and facial recognition enabled by AI has improved security in

many circumstances – but faking biometric data for identity theft and fraud brings a new threat. Deepfakes have been used to provide the video authentication to fraudulently apply for loans or enter into contracts. In 2023, a financial services company signing up new clients used a 'liveness' check involving uploading video selfies. Suspicious similarities in the backgrounds to the videos, with other clues, alerted security checkers to a host of deepfakes being used by a criminal organization in Eastern Europe to create fraudulent accounts.

Fraud can be even more audacious, working in real time. In 2019, the chief executive of an energy firm in the UK was tricked into transferring €220,000 to scammers by a call impersonating his boss's voice. And a French finance worker transferred $25 million to fraudsters who used a deepfake of his company's finance officer in a video call in 2024.

Not all AI use in crime relates to deepfakes. It can be used to target malware, ransomware, and phishing attacks more precisely, in much the same way that offers on Amazon or a social media feed are targeted to individuals' interests and vulnerabilities. AI systems can make malware more effective, and bypass or disrupt anti-malware and facial recognition systems. AI can tweak and refine the messages used in phishing and other attacks to find the most successful phrasing and tactics.

AI can also streamline and spread the information needed to commit crimes. Mainstream chatbot systems, like ChatGPT, are banned from providing information that would help someone commit a terrorist attack or other crime. ChatGPT won't tell you how to make a pipe bomb, for instance. But unregulated chatbots can provide any kind of information. This not only helps terrorists to plot and perform an attack, but makes it difficult for law enforcement to arrest and prosecute them. Possessing the instructions for making a bomb is a crime, and can be used as evidence in a trial. Just having access to a chatbot that could tell you how to do it is not an offence. Using AI for this helps terrorists cover their tracks.

Good AIs can be put to bad use. In 2022, an international security conference invited a team of biochemical researchers to show how AI techniques used to design new drugs could be abused. After training on the molecular structure of potent toxins such as the nerve agent Novichok, their AI was able to produce 40,000 molecular structures for possible biological weapons in just six hours. Some of these matched existing known chemical warfare agents, and some were more toxic

than any currently known. Others were in an entirely different class of molecules from any the training data included. The model, which is usually used to determine toxicity before beginning to develop and test a drug, can easily be repurposed for harm. The team, chastened by their experiment, called for protections and guidelines that have not yet appeared. Terrorists could manufacture a new pathogen and either release it or hold the world to ransom. It's not a meaningless threat. Some people anticipate (sometimes gleefully) the end of the world because of their religious beliefs. Others just want to wipe out a particular national or ethnic group. As we saw with Covid-19, it's impossible to contain a contagious disease in the interconnected modern world.

Killer robots have been a staple of sci-fi movies and stories for decades. Now they are becoming a real threat, though not in quite the guise we are used to. In reality, killer robots are most likely to be employed by the military to kill particular enemies. For terrorists and lone wolves, the possibility of weaponizing self-driving vehicles is more immediately accessible. Setting a car or truck to drive into a crowd no longer requires a suicide-driver.

Lethal autonomous weapon systems (LAWS) are – or would be – automated systems to pick out, pursue, and

In Ukraine, drones are used extensively to attack personnel, vehicles and sites.

kill human targets with little or no human oversight. A Russian drone with the un-snappy name Kalashnikov ZALA Aero KUB-BLA that has apparently been used in both Syria and Ukraine can work from an uploaded image to find and recognize a target. It can possibly act either with direction or autonomously, though there is so far no evidence that it has been used in autonomous mode – that is, finding and then killing a target without oversight. The demilitarized zone between North and South Korea is patrolled by the SGR-A1 sentry bot that can launch grenades and use a machine gun. Again, it's claimed that it's only used with human oversight – for now. The SGR-A1 uses infrared and visible light vision systems, pattern recognition to distinguish between humans and animals, and can identify and track targets up to 4 km (2.5 miles) away.

It seems unlikely that states with a stake in LAWS will be keen to outlaw them, so it's a threat we are stuck with.

For law enforcement agencies, the rising threat of AI-assisted crime demands AI-assisted crime-fighting. AI emergency call-handling systems can transcribe calls as they happen and immediately search police databases, cross-referencing details with firearms licences, details of previous offences, and unsolved crimes. An AI system can triage calls, helping to deploy officers to the most urgent cases. AI can also use footage from CCTV cameras and video doorbells to help trace the path of a suspect, or hook into numberplate recognition systems to find a moving vehicle.

Just as AI can be used to make deepfakes, so it can be used to reveal them. Microsoft's Video Authenticator will analyse a video and score the chances of it being fake or manipulated. Early on, unnatural blink rates and poor synchronization of lips with speech were often a tell-tale sign of fakery, but deepfakes have improved beyond those errors. Now, AI scans images for inconsistencies so tiny the human eye doesn't spot them: give-away colouring at a pixel-by-pixel level, or tiny but unrealistic variations in audio tone, for instance. Another approach is to examine evidence of liveness, such as looking at the pattern of blood flow under the skin. Even an indication that the other person on the video call might be fake should be sufficient to alert us and reduce the chance of being scammed. This might help with fraud attempts but it's no consolation to the people targeted by malicious porn. At present, criminals seem to have the upper hand as AI-assisted crime is very hard to prevent.

CHAPTER 12

FLYING BLIND

Streets filled with driverless cars is one way in which AI might change the world in the near future.

AI presents both threats and opportunities. On the one hand, it has accelerated important scientific discoveries and made possible tasks that were previously inconceivable. On the other, it is rewriting the geopolitical landscape, passing important powers to unelected oligarchs, flooding the internet with disinformation and fake pornography, and taking a toll on the environment. To chart a path that makes the most of AI and avoids the worst possible outcomes will take a concerted effort, regulation, and international cooperation.

In 2000, we had no Facebook, no iPhone, no web 2.0 and scant, primitive AI. In 1975, we had no personal computers of any recognizable type, no mobile phones, no world wide web. And in neither 2000 nor 1975 could anyone have predicted what would be in place 25 years later. Sci-fi films showed robotic AIs that either went rogue and tried to rule or destroy humanity or, as in Steven Spielberg's film *A.I. Artificial Intelligence*, are conscious and can be emotionally exploited. People proposed we might never have to work, or only for a few hours a week, because intelligent robots would do all the work. As early as 1930, economist John

Keynes predicted that his grandchildren would only work 15 hours a week.

Today we have sophisticated AI. It hasn't slaughtered or enslaved us, it's not conscious or even close to being conscious, and we are working as many hours as ever, often more. But for how long?

We like to imagine we can anticipate what the world will bring in 2050, but we are unlikely to be any more accurate than our forebears. With some confidence, we can predict climate catastrophe, mass migration of refugees, further wars, even deeper financial inequality, dictatorships replacing democracies, and AI used everywhere for surveillance and oppression. But we might be wrong. By 2050 we might have found a great source of clean energy, have liberated people from terrible jobs, stalled climate change, found the key to improving health and be feeding an equitable and peaceful world (all beneath a sky dotted with flying pigs). If we're still here, our future will probably be somewhere between these two extreme scenarios. But however it goes, AI will inevitably have played a large role in getting us there.

Some things we can be fairly sure will come. More self-driving vehicles are a near-certainty, and a corresponding reduction in the jobs for professional drivers. The current road

systems are designed for human-driven vehicles, but if more (or all) are driven autonomously, the layouts can change. Vehicles will probably be able to communicate with each other using transponders and negotiate rights of way at junctions, perhaps removing the need for traffic signals and eliminating congestion and snarl-ups. More healthcare diagnostics are likely to be AI-assisted, perhaps cutting waiting times, making tests more reliable so providing early diagnosis, and improving preventive care – and hence outcomes. The flood of disinformation will become worse, further undermining democracy and the perceived value of any information, but there will probably also be more attempts to verify true information and flag false content. AI will be turned against itself in identifying fakes and catching out cheats by flagging student papers written by AI. AI should bring greater efficiency in many industries and services, but it's likely to be accompanied by many more cases of injustice stemming from decisions made by AI without oversight or right of appeal.

Other outcomes are less certain at this stage. AI could remove so many jobs that the world economy is destabilized, maybe leaving a large body of possibly demotivated, angry people with nothing better to do than plot revolutions. AIs gambling against each other on the stock market could crash markets, or wipe out the investments of individuals – or just concentrate the world's wealth in the hands of the few with the best AI. On the positive side, robotic AIs could take over the most dangerous tasks. AI's proficiency at handling language might enable us to decode animal languages. Imagine really knowing what your pet is saying, or understanding the sounds whales and dolphins make.

The pace of change that AI can bring is so rapid that we will struggle to keep up with it. In particular, regulation and oversight are being left behind as AI developers go in any direction they choose, entering an uncharted cyber hinterland with no guardrails. But guardrails are needed to protect us as individuals, as coherent communities and societies and, potentially, as the entire human race.

Even the people involved in developing AI have warned about the pace of change. In 2023, a group of industry leaders issued a 'pause letter' calling for a moratorium on further AI development for at least six months to allow time for reflection, assessment, and planning. This wasn't written by techno-refuseniks, but by the very people who are developing AI:

'AI systems with human-competitive intelligence can pose profound risks to society and humanity...recent months have seen AI labs locked in an out-of-control race to develop and

deploy ever more powerful digital minds that no one – not even their creators – can understand, predict, or reliably control.

'... Should we let machines flood our information channels with propaganda and untruth? Should we automate away all the jobs, including the fulfilling ones? Should we develop nonhuman minds that might eventually outnumber, outsmart, obsolete and replace us? Should we risk loss of control of our civilization? Such decisions must not be delegated to unelected tech leaders. Powerful AI systems should be developed only once we are confident that their effects will be positive and their risks will be manageable. This confidence must be well justified and increase with the magnitude of a system's potential effects.'

Published on the Future of Life website, its signatories include Elon Musk and Steve Wozniak (a founder of Apple). The letter is still collecting signatures and had around 34,000 at the start of 2025.

Concern extends far beyond the AI industry itself. The World Economic Forum declared in 2024 that: 'In addition to the spread of false information, risks include the mass loss of jobs, the weaponization of AI for military use, criminal use of AI to mount cyberattacks and inherent bias in AI systems being used by businesses and nation-states.'

It's impossible to overstate the risks. They threaten the social fabric, as AI transforms the landscape of work and employment. They threaten geopolitics as AI can be used to manipulate and attack nations and ethnic groups and foment discord. AI threatens the environment and worsens the climate crisis, as AI is catastrophically resource-hungry. AI threatens the mental health and well-being of individuals, as society and interactions between people change and truth is lost beneath waves of disinformation and distrust. And that's just the risks with the level of AI we have already, before we get on to the dystopic predictions of what even more powerful AIs might be capable of doing.

Throughout this book, we have focused on the pattern-learning abilities that underpin generative AI and the deep learning systems that are put to work on problems such as protein folding, translating between languages, and driving autonomous vehicles. But the AI industry has something greater in its sights. This is Artificial General Intelligence (AGI). If it can be created, it will approximate, or perhaps surpass, human biological intelligence by bringing into play aspects that current AI lacks, such as genuine creativity, understanding, reasoning, contextual awareness, and

perhaps even empathy. A recent survey of when industry leaders thought AGI might be achieved gave responses ranging from 2027 to the next century.

An AGI should, many people think, be capable of improving itself, or designing a better AGI. This could send it into a loop of exponential improvement, giving us a super-intelligent AI in just a matter of months or years. The point at which AI becomes more intelligent than humans is known as 'the singularity' in geek circles. This might have more intelligence than the whole of human civilization. And where will that leave us? No one knows. Perhaps as gods, perhaps as slaves, perhaps annihilated.

In 2023, Geoffrey Hinton, a leading pioneer known as the 'godfather of AI', resigned from his position at Google so that he could speak openly about the threats posed by AI. Hinton won the Nobel Prize for physics for his work developing the artificial neural networks used in AI. In 2024, he declared that the chance of AI leading to human extinction in the next 30 years was now 10–20 per cent. The Asilomar AI Principles state that 'Advanced AI could represent a profound change in the history of life on Earth, and should be planned for and managed with commensurate care and resources', yet we are doing scant planning.

One rarely mentioned impact of AI is its environmental footprint. The drain on resources of AI is immense. Whatever else the future of AI brings us, environmental disruption is a certainty.

The large volume of computer equipment needed to train and run AI draws on scarce minerals and rare metals that are generally not sustainably mined. To make a 2 kg (4½ lb) computer uses 800 kg (1,764 lb) of resources, and a lot becomes toxic waste at the end of its life. The huge data centres AI needs use vast amounts of energy and water to run and cool their computers. Around five per cent of the USA's total electricity use is by AI data centres. In the Republic of Ireland, it's likely that a third of all electricity will be used by data centres in 2026. The National Grid in the UK predicts the demand for electricity by data centres will rise six-fold over the ten years to 2034, driven largely by AI. A search carried out by ChatGPT uses around ten times as much electricity as the same search carried out by a regular search engine without AI. To train a robotic hand with AI to solve a Rubik's cube used as much power as the three nuclear power stations output in an hour.

The demand for data centre acreage is expected to nearly double over the four years from 2023 to 2027. If AI grows as quickly as expected, its water use by 2027 will be four to six

The Middenmeer data centre in the Netherlands.

times that of Denmark. Around 15 simple ChatGPT queries use around 500 ml (16 oz) of water to process – and there are billions every day.

There is a move to make some AI more sustainable. Microsoft claims to have reduced wastewater by 95 per cent and aims to be 'water positive' by 2030. The data centre construction company Iron Mountain has built its centre outside Pittsburgh 60 m (200 ft) underground so that it is naturally cooler, and draws its water from a lake in a disused quarry, a source not used by the local community.

These are small glimmers of hope, but the eco-footprint of AI for now is disastrous. When AI companies, such as Google, aim to use renewable energy to run their data centres, it looks like greenwashing. We don't have enough renewable energy to replace fossil fuel use, so squandering what we do have on making videos of dancing yaks and memes isn't a smart move. It just leaves other consumers using energy derived from fossil fuels. DeepSeek costs less in cash and resources to develop than OpenAI's products. It probably used an order of magnitude less energy and water. But paradoxically, its success might only increase the eco-footprint of AI in the long run. If it makes AI more attractive and accessible, use will increase.

Can we look to AI for a solution to the climate crisis, to cancer, to world hunger? Currently, no. These are problems that involve data in many different, often unconnected, fields – and political will. To get to the root of 'what causes cancer' an AI would need the full details of the lives and bodies of a huge number of cancer patients and people without cancer – and

that means *every* detail including all pollutants they have been exposed to and the diseases they have had and their genome and what they have eaten, and so on. Even if AI is tracking you – and it is – it's a long way off having all that information.

Like plastic, AI is good and bad depending on how it's used. The oceans are filled with plastic waste and the infosphere is filled with disinformation, porn, and nonsense. The flippant and clumsy use of generative AI does more harm than good, but some of the products of AI in science and medicine have brought world-changing benefits. If humanity is to benefit from AI, we need to get over seeing it as a new toy and integrate it thoughtfully into societies. That means outlawing iniquitous and exploitative uses, regulating it so that it is used as a tool by people rather than as a means to replace or control people, and eradicating the bias inherent in it now.

AI can be used to our detriment or our benefit. It's up to us as voters and consumers to decide how much harm we will allow and cooperate with, and how much power we will hand to the unelected tech-bros and the governments that deploy AI. We need to see beyond the novelty value and work out how to use AI safely and sensibly, before it's too late.

INDEX

PICTURE CREDITS